Statistics and Probability

Statistics and Probability

Second Edition

S. E. Hodge
M. L. Seed

Blackie
Chambers

Blackie & Son Limited   Bishopbriggs, Glasgow G64 2NZ
14–18 High Holborn, London WC1V 6BX

W. & R. Chambers Limited   43–5 Annandale Street, Edinburgh EH7 4AZ

First published   1972   Reprinted 1974, 1975
Second Edition 1977
Latest reprint 1985

Blackie   0 216 90450 1

Chambers   0 550 75998 0

Printed in Great Britain by Thomson Litho Ltd, East Kilbride, Scotland

Preface

Our intention in this book is to present a development of basic probability and statistical theory which is both readable and relevant. Most sections start with an everyday problem discussed initially at an intuitive level. Mathematical concepts are then introduced gradually and naturally, with the aim of building a mathematical model which is simple and yet approximates to the real-life situation.

We have nowhere tried to avoid the mathematical foundations of the subject but, in facing up to the more technical points, we have taken care to explain the terms and techniques and to derive the formulae we require where it is possible to do so.

Use has been made of modern mathematical concepts and terminology (e.g. sets, matrices, calculus) where this makes the ideas clearer.

Some of the earlier work may have been covered in previous courses but it is included here for the sake of completeness.

The second edition contains a number of changes. These mainly affect the early chapters where the approach to probability has been modified to take account of recent developments.

The book is particularly suitable for both English sixth-formers and Scottish sixth-year students since the material covers most of the syllabuses for the major English A-level examinations which include statistics, and also the Scottish Certificate of Sixth Year Studies. It could in addition be used as the basis of a "liberal studies" course for sixth forms.

College, Polytechnic and University students who require some basic statistics should find that this text covers much of the material they require.

We have included a wide selection of examples, some involving actual data collection and some extending the theoretical work in the text. A revision exercise is included after each section of chapters. We are grateful to the Oxford and Cambridge Schools Examination Board, the

Cambridge Local Examinations Syndicate and the Joint Matriculation Board for permission to include many of their questions. The source of individual questions is indicated by the appropriate abbreviations.

We are indebted to the Football League for permission to reproduce the table on page 214, and to Cambridge University Press for portions of tables from Lindley and Miller's *Cambridge Elementary Statistical Tables*.

We acknowledge with thanks the help of the sixth-formers at the Cavendish School, Hemel Hempstead, on whom many of these ideas were first tried.

Thanks are also due to Malcomn Marshall of the University of Kent and to various relatives for their help in working examples, typing and proof-reading.

Contents

Contents

The Meaning
of Probability

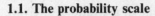

1.1. The probability scale

(V) Arsenal will win their next match against Liverpool.
(W) I shall never be a millionaire.
(X) The work known as Beethoven's *Fifth Symphony* was not in fact written by Beethoven.
(Y) A die will land with the four uppermost.
(Z) A British man chosen at random would vote Labour if there were an election next week.

Try arranging the above statements in order of likelihood. Now place each event on a scale ranging from *completely impossible* to *absolutely certain*.

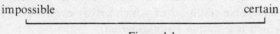

impossible   certain

Figure 1.1

There will probably be a considerable measure of agreement about the positions of events W, X and Y, even if the placings of V and Z are more open to personal prejudice; so it does look as though we all have an intuitive idea of a scale of probability. The next step, therefore, is to try to quantify the concept.

We start by attempting to assign a numerical value, to be called the *probability,* to each point on the scale. As on a temperature scale we will have to begin at the two fixed points, in this case the extremes of impossibility and certainty. It would seem reasonable to say that an impossible event has zero probability of occurring. The other extreme is perhaps more arbitrary (we could choose 100 for instance as in the Celsius scale), but it turns out to be convenient to say that an event which is certain has probability 1. The probability of any event is then a number lying between 0 and 1.

1

1.2. Dice: symmetry method

An event with a simple probability is "a fair coin will fall heads". By saying the coin is *fair*, we mean that we are assuming that the coin is equally likely to fall head uppermost or tail uppermost. and therefore that the event "falling heads" lies exactly half way along the probability scale.

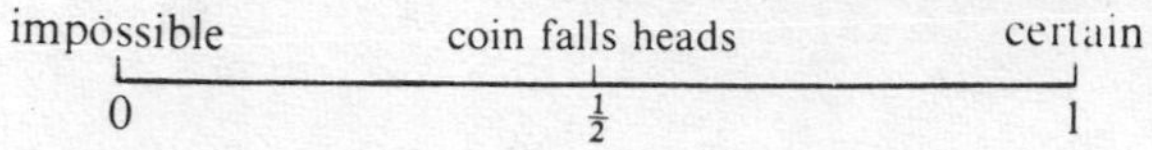

impossible   coin falls heads   certain

0                                  $\frac{1}{2}$ 1

Figure 1.2

It should likewise be easy to place the event Y—that a die will land with the four uppermost—provided that we assume that the die is fair. A little thought shows that the event Y is 5 times as likely not to happen as to happen, and must therefore divide the scale in the ratio 1:5, giving it a probability of $\frac{1}{6}$.

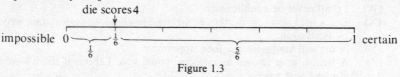

die scores 4

impossible 0 —— $\frac{1}{6}$ ———————————————— 1 certain
                $\frac{1}{6}$ $\frac{5}{6}$

Figure 1.3

(As in betting terms, odds of 5:1 against are equivalent to chances of 1 in 6 and therefore a probability of $\frac{1}{6}$.)

Suppose we were now to ask what the probability would be of scoring either a three or a five with a fair die. We shall refer to this event as event A and use $P(A)$ as shorthand for *the probability of event* A. Clearly we again have 6 possible results, 2 of which satisfy event A, so $P(A) = \frac{2}{6} = \frac{1}{3}$ (see Fig. 1.4, which is a simple *Venn diagram*).

Would the result be the same if the die were biased; for instance if a corner were chipped off? Probably not, as it might make the chance of landing on one face less than that of landing on another. So it is only when we feel justified in assuming that all the possible outcomes of an "experiment" are equally likely that we can work out a value for the probability in this way, which in practice poses a severe restriction.

Can we generalize our method of finding the probability of event A? We started by considering an "experiment", namely, tossing a die, which had a set, which we shall call the outcome space, E, of possible outcomes which we felt justified in assuming to be equally likely. Here $E = \{1, 2, 3, 4, 5, 6\}$ and $n(E) = 6$, where $n(E)$ is the number of outcomes in E. The event A concerned with the outcome of the experiment was satisfied by a subset of E, which we shall also refer to as A. Here $A = \{3, 5\}$, and $n(A) = 2$. We then calculated that

$$P(A) = n(A)/n(E) = \frac{2}{6} = \frac{1}{3}.$$

So one way of arriving at a numerical value for a *probability is:*

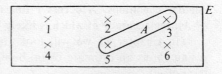

Figure 1.4

If E is the set of outcomes of an experiment, all of which are assumed to be *equally likely,* **and an event A is satisfied by a subset A of them, then** $P(A) = n(A)/n(E)$.

This method, sometimes called the theoretical method, can only be used when there is a physical symmetry about the situation which results in no member of the outcome space being inherently more or less favoured than any other. Of course because, for instance, no *real* die will be completely symmetric, the symmetry method can only give an estimate of the probability in any particular case.

1.3. Biased die: frequency method

If we return to the event Y, that a die will land with the four uppermost, would we be able to estimate $P(Y)$ if we knew the die to be biased? The symmetry method is of little use to us here since the assumption that each face is equally likely can no longer be justified. We must return to what we really mean by probability.

For instance, if we threw a fair die 60 times, we would expect roughly 10 scores of four, though we would not be too surprised if it varied between 8 and 12, say, which is an inaccuracy of $\pm 20\%$. It could even happen that we got 60 scores of four, though this is highly improbable, so we cannot pretend to predict the results very accurately. However, if we threw the die 6000 times we would expect about 1000 scores of four and probably only a variation of 20–30 on each side; this brings the inaccuracy down to 2–3%. In other words, the greater the number of times we repeat an experiment, the more accurate we expect our probability value to be as a prediction of the results.

This suggests a method of finding the probability of our biased die scoring four. For instance, if we throw it 1000 times and obtain 307 fours, we would expect the probability to be somewhere near 0·307. Of course this would only be an estimate, but the greater the number of times that the experiment is repeated, the more accurate we would expect the estimate to be. The graph on page 4 shows what happened to the proportion of fours obtained when a die was thrown 140 times.

This gives an alternative way of estimating probability, which we shall call the *frequency,* or the *experimental,* method; namely:

If an event A is satisfied by a certain outcome (or outcomes) of an experiment and, when the experiment is repeated *n* **times under exactly**

similar conditions, A occurs *r* times out of *n*, then r/n is an *estimate* of $P(A)$, the accuracy of which is likely to increase as *n* increases.

(In Chapter 11 we will look further at the likely error involved in using r/n as an estimate of probability, and at how this depends on the size of *n*.)

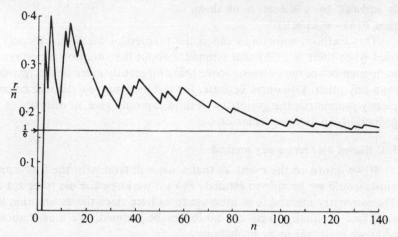

Figure 1.5. Experiment graph of r/n as *n* increases for the event of scoring 4 on a die.

1.4. Football results: subjective method

"Arsenal could either win, draw or lose their next match against Liverpool, so the probability of their winning is 1/3." What is the flaw in this argument?

In fact the probability of a win for Arsenal cannot be estimated on grounds of symmetry, since the alternative outcomes cannot reasonably be regarded as equally likely; nor is it possible to obtain a good estimate based on the frequency method, as a particular match cannot be repeated a large number of times under closely similar circumstances. In such situations if it was necessary to arrive at a reasonable estimate, one might choose to gather a variety of evidence based on expert opinion, results of recent games, team form and so on, and to arrive at an acceptable estimate by a weighing of all this evidence. Such a method would for obvious reasons be known as *subjective* since faced with the same evidence two people may well differ fundamentally in their estimates. (Two people doing their own experiment to arrive at a probability value by the frequency method would also be very likely to get different results even using the same number of repetitions. However, given the same experimental data, they would arrive at the same estimate.)

The subjective method is therefore more open to error than the other two methods, but there are many important real-life situations in which no alternative method is available. Two such instances are that of a managing director in an industrial firm who in order to decide whether to invest in new plant must estimate the probability of a recession in the period when it first comes on line or of an insurance broker who has to work out the probability that a new type of tanker will run aground.

1.5. Election results: sampling methods

Let us now return to event Z in our original list, namely that a man chosen at random would vote Labour if there were an election next week.

If you had been asked to find the probability that a man chosen at random from your road would vote Labour, it should not present too many difficulties; if there are 50 men in the road and on inquiring you are satisfied that 31 will vote Labour, then the chances of any one chosen at random voting Labour are 31 in 50, so the probability is $\frac{31}{50} = 0.62$. This is really just a case of the symmetry method, with E = {men in your road}. The application is valid, as the very phrase *chosen at random* must mean that each man has an equal chance of selection. We assume for simplicity that no-one is undecided or changes his mind.

When we widen our population to include the whole country, mathematically speaking it does not alter the problem very much. The difficulty this time is not in the mathematics but in the question of how to evaluate the total number of male Labour voters. Our only feasible solution is to select some sample of the population as in a Gallup Poll, and use the probability found from this as an estimate of $P(Z)$. As in the case of the frequency method, the larger our sample the more accurate will our estimate be, but the greater the cost of the survey. A compromise between accuracy and economy will be necessary.

How do we pick our sample? Most sampling methods are a mixture of random sampling and stratified sampling.

Random sampling

A random sample is one in which each member of the total population has an equal chance of being selected. (We use *population* in the statistical sense to include any collection of things or people.)

A well-known example of random sampling is that of ERNIE, an electronic method of producing random numbers to identify winning premium bonds. A simpler method is to use a table of random numbers (there is one on page 246). If a number is assigned to each member of the population, then a random sample can be selected by beginning at an arbitrary place in a random number table and identifying the members of the population corresponding to the numbers as they appear. You will probably be able to work out suitably economic ways of using the tables when the numbers go from 1 to 200 only, or contain a letter as well. Of course a roulette wheel or a 20-sided die (try making a 10-sided one!) would be just as efficient as generators of random numbers. Equally effective for small populations, if less sophisticated, is the well-tried method of drawing names or numbers from a hat.

A non-random method which in practice gives similar results is to select every tenth person off a voting list, school register or list of householders.

Great care must be taken to avoid bias in the method of sampling; otherwise the most unlikely results can appear, as the following stories show.

(i) A researcher was extremely worried by the fact that a "random" sample of people in the street showed a huge proportion of regular gamblers on horse-racing until he realized that the street he had chosen was near Liverpool Street station, from which a train was about to leave for Newmarket.

(ii) You may also be able to suggest why an inquiry involving choosing random names out of the telephone book and lists of car owners many years ago wrongly predicted that a Conservative Republican would win the Presidency of the United States.

Random sampling is a useful technique giving a reasonable estimate for the total population, provided that the sample is large enough. However, in a population the size of the British one, identifying and contacting those selected would be relatively difficult.

Stratified sampling

When a stratified sample is required, the population is first divided into groups or strata according to one or more criteria, such as sex, age, social class, or income. The sample is then formed by choosing a sub-sample at random from each group and combining the sub-samples. The numbers in each sub-sample are usually chosen so as to make the proportion of the sample in each group approximately the same as the proportion in the population. For example, university students could be stratified according to faculty, and the sub-samples chosen so that the proportions of Arts, Science and Engineering students in the sample are the same as in the whole university. Alternatively the sub-samples from each faculty could be of equal sizes, but the results weighted according to the relative proportion of the faculties.

Stratified samples usually give more accurate estimates than simple random samples, and can also be easily used to obtain separate estimates for the strata.

Often two or more sampling schemes are combined. In most Gallup Polls a sample of constituencies is first chosen and then a random sample of voters picked from each of the chosen constituencies. This is known as multi-stage sampling; another example would be in choosing classes of 12-year-old children where a random sample of local education authorities might first be selected, in each of which a sample of schools is randomly picked, and in each school a class of 12-year-olds are randomly chosen.

Quota sampling

This method is related to stratified sampling but is used when a small but representative sample is required, for example, where 50 people are required to take part as a panel in a television programme. The sample might be required to match the adult population in including, say about 55% of females, 8% of professional workers, 15% of people over 60, and so on. The first person chosen might not satisfy all these criteria, and the researcher would then try deliberately to choose the next person in the sample so as to fill in some of the remaining categories, hoping at the end to have a list which at least approximately matches the requirements.

1.6. The nature of probability

Although we have now given four methods of estimating the probability of an event, we have not actually defined what we mean by the probability of an event, except that it is a number between 0 and 1 inclusive which is assigned to the event, on some rational grounds, to indicate how likely we feel it is to occur.

In fact it is no easier to define "probability" than it is to define properties like "mass" or "force"; in all such cases it is enough to have an intuitive feeling for the nature of these properties, to know how to measure them, and to know what formal rules (like $F = ma$) they obey. An example of one probability rule is given below; others follow in Chapters 2 and 3.

1.7. The probability that an event does not happen

If the probability of a fair die scoring 4 is $\frac{1}{6}$, what is the probability of it not scoring 4? Clearly it is $\frac{5}{6}$; because, as the event happens in one outcome out of six, there are five outcomes out of six when it does not. Similarly if experiment shows that the probability of my getting to school on time in the mornings is $\frac{2}{5}$, then it would seem reasonable that the probability of my being late is $\frac{3}{5}$.

In general, if A is an event and we denote by A' the event that A does not happen, then it looks as though $P(A') = 1 - P(A)$. We accept this as one of the "rules" that all probabilities obey.

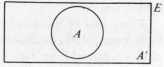

Figure 1.6

INVESTIGATIONS I

In this exercise, where you obtain a probability experimentally, describe the method as well as the results. Also say how accurate you think your answer is, giving reasons and suggestions for improved methods.

1. Discuss suitable methods of choosing a sample from each of the following populations:
   (a) pupils at a school,
   (b) packets of a certain brand of soap powder after distribution to shops,
   (c) trees in a wood,
   (d) a lorry load of potatoes,
   (e) the population of England,
   (f) worms in a field.
2. If you pick at random a number between 1 and 10, calculate the probability that it is: (a) seven, (b) even, (c) a perfect square.
   Test your results experimentally by:
   (i) asking a sample of people to select a number at random,
   (ii) using a pack of cards,
   (iii) using the table of random numbers on page 246.
   Account for any discrepancies between the results.
3. If you were to pick at random a football match in one of the first four Divisions, estimate the probability that it would result in (a) a home win, (b) an away win, (c) a total score of more than 4 goals. As far as you can tell, are the probabilities any different for Scottish football?
4. Estimate the probability that a British citizen, taken at random, (a) is left-handed, (b) wears glasses, (c) was born in March, (d) has exactly one brother or sister, (e) prefers wine to beer, (f) is male, (g) is aged 25–34, (h) earns more than £3000 a year, (i) is a married man, (j) is both (g) and (i); or choose any other characteristics you wish to investigate. (You may find useful a copy of the *Annual Abstract of Statistics* (H.M.S.O.), which is usually available in the reference section of a public library.)
5. Estimate experimentally the probability that if you pick at random two of a set of dominoes they could be placed next to each other in a game. Can you check your answer theoretically?
6. Find the probability that, where there are three children in a family, they are all of the same sex:
   (a) experimentally using 3 pennies to "simulate" the situation,
   (b) theoretically,
   (c) experimentally, provided you can find a large enough sample.

7. Guess what you think might be the probability of two people having the same birthday in a class of 30. Check experimentally how close your answer is. (You may find school registers helpful.)

8. Estimate the probability of a dropped match falling across one of a single set of parallel lines ruled on a large sheet of paper, the distance between the lines being double the length of the match. You may notice something rather special about the reciprocal of your answer.

9. Estimate the probability that a driver taken at random is (a) female, (b) wearing a safety belt. Or estimate the probability that a vehicle taken at random will be (a) of a certain make (choose whichever you like), (b) commercial, (c) less than 2 years old. Or choose any other characteristic you wish to investigate.

10. Estimate the probability that some of the following objects land in each of the possible alternative positions when they are dropped at random:
   (a) a drawing-pin,   (b) a tin-tack,   (c) a clothes-peg,
   (d) a fork,   (e) a teaspoon.

11. Guess what you think the probability might be of being dealt a Bridge hand which (a) contains the king and ace of at least one suit, (b) contains only 3 suits. Check your results experimentally.

12. Find experimentally the probability of a total score of 5 with two dice. Check your result theoretically.

13. Arrange for someone to put 10 beads or counters in a closed matchbox or bag, so that a certain number, unknown to you, are of one particular colour. By drawing out a single item many times and replacing it immediately, estimate the probability of drawing one of that colour and hence the total number of that colour in the container (before you check by opening it).

14. Armed with a list of runners for a horse race in which there are only a small number of horses, preferably 5 or less, use a sample of people to estimate experimentally the probability that a person at random would (a) pick the winner, (b) pick out the horses which eventually come first, second and third, though not in any particular order, (c) forecast correctly the first 3 horses in the correct order. Try to work out the theoretical probabilities and comment on any discrepancies.

15. Find the probability that a coin falls completely inside a square when it lands on a piece of paper marked out in squares (as in roll-a-penny). Repeat this for different sizes of coins and squares to see whether you can guess any relationship between the probability, the diameter d of the coin and the side w of the square. Can you think of a way of working out the theoretical probability in the general case?

Answers. No answers are given to these exercises as it is intended that the results should be obtained, or checked, experimentally and any discrepancies should provoke discussion.

Calculating Probabilities

2.1. Introduction

In Chapter 1 we decided that in certain cases we could work out the probability of an event simply by considering the set of all the possible outcomes of an experiment. We thus arrived at the symmetry method of assigning probability:

If E is the set of outcomes of an experiment, all of which are assumed to be *equally likely*, **and an event A is satisfied by a subset A of them, then** $P(A) = n(A)/n(E)$.

So far we have only applied this method to finding the probability of scores on dice. These were fairly simple problems, but often the calculation of a theoretical probability requires a little more thought. For example:

George wants to see the film at the local cinema, but he cannot decide whether to take one of the three girls from his office, Heather, Tessa or Jane, or whether to go on his own (which would be cheaper). He finally tosses the penny in his pocket three times, having decided that if it lands heads three times he will ask Heather; if heads twice he will ask Tessa; if once, Jane. If they are all tails he will go on his own. What is the probability that he asks Jane?

At first glance we are tempted to say $\frac{1}{4}$, "as there are 4 possible results"; but are we sure that all the results are equally likely? In fact we have to go back to the coin-tossing and actually write out every possible outcome of the experiment. We soon realize that a run of 2 heads followed by a tail (HHT) is a different sequence of events from that of a tail followed by 2 heads (THH), although the two give the same number of heads and tails. The total set of outcomes is thus:

$$E = \left\{ \begin{array}{cccc} & \text{HHT} & \text{TTH} & \\ \text{HHH,} & \text{HTH,} & \text{THT,} & \text{TTT} \\ & \text{THH} & \text{HTT} & \end{array} \right\}$$

Heather   Tessa   Jane   alone

It seems intuitively likely that each of the eight outcomes is equally probable. Hence $P(\text{Jane}) = n(\text{Jane})/n(\text{E}) = \frac{3}{8}$.

2.2. Arrangements (or permutations)

George has more money than sense! In order to resolve his quandary about his visit to the cinema he decides to take all three girls. Assuming that the seating order is completely random, what is the probability that he will sit next to Jane?

We could write down the set E of all the possible arrangements of the four (which are assumed to be random and hence equally likely) and then extract the subset F of those in which Jane and George are next to each other, as follows:

$$E = \{\text{GHTJ, HTJG, TJGH, ...}\}$$
$$F = \{\quad\quad\ \text{HTJG, TJGH, ...}\}$$

Then, hoping that we have not left any out, we could count them up and calculate $P(\text{F})$ as $n(\text{F})/n(\text{E})$.

But surely there is an easier way of working out the number of arrangements of the four letters G, H, T and J, which gives us $n(\text{E})$, without writing them all out? Suppose we reason as follows:

There are 4 ways of filling the first seat. Let us call it seat 1, as any of the four could sit there. Once seat 1 is occupied, three people are left, one of whom must sit in seat 2. Thus, as there are 3 ways of filling seat 2 for each of the 4 ways of filling the first seat, there must be 4×3 ways of filling the two seats. This may seem clearer if we write it out more fully, using the initial letters of the names. With George in seat 1, we have three arrangements GH, GT, and GJ; with Jane in seat 1 we have JG, JH, and JT, and similarly with either Heather or Tessa in the first seat, giving a total of $4 \times 3 = 12$ ways. In the same way, for each of these 12 ways there are then two people left from which to fill seat 3. For instance, if we start with GH, we could have either GHJ or GHT. This gives a total of 12×2 or $(4 \times 3) \times 2$ ways of filling the first three seats. As there is then only one way of filling seat 4, once the first three have been filled, the total number of ways of filling all four seats is $4 \times 3 \times 2 \times 1$.

It is convenient to denote the product $4 \times 3 \times 2 \times 1$ by the expression 4!. Then the number of ways of arranging 4 people in 4 seats or, more generally, 4 objects in 4 places, is 4!.

Similarly the number of ways of arranging 7 objects in 7 places is $7 \times 6 \times 5 \times 4 \times 3 \times 2 \times 1 = 7!$.

In general, the number of ways of arranging n objects in n places is $n \times (n-1) \times (n-2) \times \ldots \times 3 \times 2 \times 1 = n!$ and is called *factorial n*.

To return to our example, we have that the number of possible arrangements $n(E)$ is $4! = 24$. Can we now find a similar way of calculating $n(F)$? As George and Jane must now sit together, let us consider the subset of E containing the combination (GJ); for instance (GJ)TH and T(GJ)H will be members of it. In fact as G and J are now "inseparable", we can regard them as one element. Hence the number of elements in this subset is the same as the number of ways of arranging the 3 elements H, T and (GJ), which is $3! = 3 \times 2 \times 1 = 6$.

Also there must be an equal number of arrangements containing the pair (JG), and so $n(F) = 2 \times 3! = 12$.

Therefore $P(F) = n(F)/n(E) = \frac{12}{24} = \frac{1}{2}$, so Jane can console herself that even if Tessa and Heather are going too, there is still an even chance that she will sit next to George.

Could we have used this sort of reasoning in our previous example in order to work out the total number of possible results of tossing a coin three times, where we had E = {HHT, HTH, ...}? The situation is not exactly the same as the fitting of objects into places, but we can use a similar argument. Each member of E is of the form ---, where the first blank can be filled in two different ways, namely H or T. But for either way of filling the first, we still have a choice of two ways, H or T, of filling the second. There are therefore 2×2 ways of filling the first two blanks. As we still have a double choice for the third place, there must be $2 \times 2 \times 2 = 8$ possibilities in all. This is the same result as we obtained when we wrote out all the members separately.

You will notice that the difference between this and the fitting of certain objects into places is that here, putting an H into the first blank does not stop us from using it in the second too, so that repeats are allowed; whereas fitting an object into the first place prevented us from choosing it also for the second, resulting in a decrease of one in the number of choices each time.

Let us consider one further example; if you did Investigations I, question 14 (p. 9), you may have estimated the experimental probability of picking first, second and third in a horse race with five runners. We can also work out the probability theoretically, assuming that the trio of horses is chosen randomly. If we call the horses P, Q, R, S and T, then E is the set of all possible choices for the three places, i.e. E = {PQR, RPQ, TSR ....}. If --- is a member of E, arguing as before, there are 5 ways of filling the first blank, then 4 ways of filling the second, and 3 ways for the third. We therefore have $n(E) = 5 \times 4 \times 3 = 60$, of which only 1 is the correct order. Hence the probability of picking the first three in the correct order out of five runners is only $\frac{1}{60}$. You can imagine how small it is for the Grand National!

Notice that we could have said that the number of ways of selecting, in a definite order, 3 things out of 5 is $5 \times 4 \times 3 = 5!/2!$. This leads us to a general formula:

the number of ways of selecting, in a definite order, r things out of n is

$$\frac{n!}{(n-r)!}$$

For $r = n$ we already know that the number of ways is $n!$ Substituting in our new formula to make the result consistent we must define $0!$ to be 1.

EXERCISE 2A

1. A die is in the shape of a regular tetrahedron. It has the numbers 1, 2, 3, 4 marked on its faces. After each throw the number which is not seen, i.e. the base number, is recorded. A "trial" consists of two throws of the die. Every possible result of the trial is a number pair. Plot points to represent these on a rough graph. Hence or otherwise determine the probability of each of the following:
   (a) that the sum of the numbers does not exceed 5,
   (b) that the sum is odd,
   (c) that the difference does not exceed unity,
   (d) that the numbers are the same. (S.M.P.)
2. Four girls Ann, Brenda, Carol and Diana go shopping together in a busy supermarket. They each join a different queue to pay. All the queues look the same length. Find the probabilities of (a) Ann leaving before Brenda, (b) Carol being the last to leave, (c) Ann leaving first followed by Diana, (d) Carol and Brenda both being outside before either of the others. (Assume that no two can leave at the same instant.)
3. A school blouse is made in three sizes: small, medium and large. In a class of 24 girls, 5 can wear the large size, of whom three can also wear the medium size (but not the small size). Although 18 of the class can wear the medium size, 7 can wear the small size, and all can wear at least one size.
   (i) Draw a Venn diagram to illustrate this.
   (ii) If one girl is chosen at random from the class, find the probability that she can wear:
   (a) a large-size blouse,
   (b) a medium size but not a large size,
   (c) both a medium size and a small size.
   (iii) If a girl wearing a small-size blouse is chosen, find the probability that she cannot wear a medium-size blouse. (S.M.P.)
4. An integer is chosen from the first 1000 positive integers; if each is equally likely to be selected, calculate the probability that the integer chosen is not divisible by 3 or 7. (M.E.I.)
5. Three bus services call at a bus stop; the services go to Swiss Cottage, Golders Green and Kensington respectively. The times of arrival at the stop in minutes during each hour are shown below:

Swiss Cottage	00	20	30	40	55
Golders Green	05	15	25	50	–
Kensington	10	35	45	—	—

A bus leaves the stop just as a passenger unfamiliar with the time-table arrives. Find the chance that:
  (i) the next bus is going to Kensington,
  (ii) the next bus is not going to Swiss Cottage,
  (iii) neither of the next two buses is going to Golders Green.   (Cam.)
6. The digits $1, 2 \ldots n$ where $n > 5$ appear in random order in a row. Find the probability that 4 and 5 will be adjacent.   (J.M.B.)
7. Six people are queuing for a bus. What is the probability that:
  (i) they are in order of increasing age,
  (ii) the three youngest are first, again in order of increasing age,
  (iii) the first three are in order of increasing age?
8. How many non-zero even numbers less than one million can be made from some or all of the digits $1, 4, 0, 7, 9$:
  (a) if repeats are allowed,
  (b) if repeats are not allowed?
  If one of these numbers is chosen at random, find the probability in each case that it is less than 5000.
9. George eats 7 biscuits for his tea. He likes to vary the order each day. In how many ways can he do this:
  (a) if the biscuits are all of different sorts,
  (b) if he has 2 chocolate digestives but the others are all different,
  (c) if he has 3 gingernuts, 2 chocolate digestives, 1 coconut and 1 shortbread?
  Write the answers in factorial form. Can you suggest a formula for the number of arrangements of n things of which p_1 are of one sort, p_2 of another sort, etc?
10. Six men enter a room in which 8 chairs are arranged in a straight line. If the men take seats at random, in how many different ways may they be seated? Calculate the probability that two particular men will be seated in adjoining chairs.   (S.M.P.)
11. Three dice, each with faces numbered 1 to 6, are rolled together. Find the probability that:
  (i) three different numbers will appear,
  (ii) two consecutive numbers will appear.
12. The small country of Transvania is divided into three regions, ALENTORIS, BAKKYLIVO, and CALLADIA. Every car in Transvania must carry a registration code beginning with three letters chosen from the name of the region in which it is registered. No letter may appear in any code more times than it does in the name of the corresponding region. If the letters are followed by any number from 1 to 100 inclusive, find the possible numbers of cars which can be registered in each of the three regions. What is the probability that a car carries the letters OIL in each of the three regions?
13. There are five other people with you at a dinner party. What is the probability that:
  (a) you are seated next to the hostess,
  (b) you are seated between the hostess and her husband,
  (c) the 3 women alternate with the 3 men?
14. I remember from last year that I had 3 pink, 4 blue and 2 white hyacinths, but I cannot remember the colour of any particular bulb. If I plant the bulbs in a row in a window-box, in how many ways can they be arranged? Find the probability:

   (i) that the two end flowers are white,
  (ii) that each pink flower has a blue flower on either side of it,
 (iii) that the arrangement is symmetrical,
 (iv) that the third bulb from the end is white.
15. Wall can openers are equally likely to be blue, yellow or red, and are packed in
plain boxes. A customer demands a blue one. What is the probability that the
first box opened will contain a blue one? What is the probability that the
first will not be blue but the second will? What is the probability that none of
the first four boxes will contain a blue one?

2.3. Selections (or combinations)

If the probability of our picking the first three horses in the correct
order is only $\frac{1}{60}$, how much more likely are we to pick the right three
horses, though the order may be incorrect?

Let us look at the set E again, but this time written out so that all
the results with the same horses in the first three are in the same column:

$$E = \begin{cases} \text{PQR, PQS, PQT, QRS, } \ldots \\ \text{PRQ, PSQ, PTQ, QSR, } \ldots \\ \text{QRP, } \ldots \\ \quad \ldots, \end{cases}$$

Each column consists of all the possible arrangements of each set of
three letters and so has $3! = 3 \times 2 \times 1 = 6$ elements. However each
column represents a different selection of three out of the five horses.
Therefore the probability of getting the correct three regardless of order
is $\frac{6}{60} = \frac{1}{10}$, as we saw that E had 60 members. Or we can say that as there
are six elements in each column there must be 10 columns altogether,
and each column represents a different selection of three horses. As
each selection is equally likely and only one gives the correct selection,
the probability of choosing the right trio is again $\frac{1}{10}$.

We have incidentally found that the number of ways of choosing
three horses out of five, regardless of order, is 10. We found this by
writing down the total number of ways of selecting three horses in order,
5!/2!, and then dividing this by the number of different selections which
produce the same actual trio of horses, 3!. So we see that the general
formula for the number of ways we can select r things out of n will be

given by: $\dfrac{n!}{r!(n-r)!}.$

Let us denote by $\dbinom{n}{r}$ the number of ways of selecting or choosing r things

out of n. Then we can deduce some properties of $\dbinom{n}{r}$.

(i) $$\binom{n}{n-r} = \frac{n!}{(n-r)!\,r!} = \binom{n}{r}$$

(if we choose r things out of n we "choose" the remaining $(n-r)$ things by leaving them behind).

(ii) $$\binom{n}{1} = \binom{n}{n-1} = \frac{n!}{(n-1)!\,1!} = n.$$

From our formula $\binom{n}{n} = \frac{n!}{n!\,0!}$. We agreed to define $0! = 1$ in the previous section and hence we have, as we should expect, $\binom{n}{n} =$ the number of ways of choosing all n things $= 1$.

There is an easy way of forming the values of $\binom{n}{r}$ from smaller values, for if we arrange the values in a triangle we get:

$$\binom{0}{0}$$
$$\binom{1}{0} \quad \binom{1}{1}$$
$$\binom{2}{0} \quad \binom{2}{1} \quad \binom{2}{2}$$
$$\binom{3}{0} \quad \binom{3}{1} \quad \binom{3}{2} \quad \binom{3}{3}$$
$$\binom{4}{0} \quad \binom{4}{1} \quad \binom{4}{2} \quad \binom{4}{3} \quad \binom{4}{4}$$

or

```
        1
      1   1
    1   2   1
  1   3   3   1
1   4   6   4   1
```

...   .

Figure 2.1

You may recognize this as Pascal's triangle; in any case you will be able to see how each row can be built up from the one before.

We can now see how to make use of these results in the following sort of problem:

What is the probability that a Bridge player holds all four aces in his hand? In this case E is the set of all possible hands (each of which will be equally likely if the dealing is fair) so $n(E)$ will be the number of ways of selecting 13 cards out of 52, i.e. $\binom{52}{13}$. If we now consider the subset A of hands with 4 aces, then each hand in A has 4 of its cards fixed, so there are only 9 left to be chosen from the remaining 48 in the pack. Hence $n(A) = \binom{48}{9}$.

Therefore $P(A) = \binom{48}{9} / \binom{52}{13} = \frac{48!}{9!\,39!} \Big/ \frac{52!}{39!\,13!} = \frac{13.\,12.\,11.\,10}{52.\,51.\,50.\,49} \simeq 0{\cdot}00265.$
In other words you are only likely to enjoy playing this sort of hand once in about 400 hands.

2.4. The addition law

In Chapter 1 we considered briefly the probability of scoring three or five with a fair die. We had $E = \{1, 2, 3, 4, 5, 6\}$, $A = \{3, 5\}$ and hence $P(A) = \frac{1}{3}$. Putting this another way, $P(3 \text{ or } 5) = \frac{1}{3}$. But $P(3) = P(5) = \frac{1}{6}$, so that $P(3 \text{ or } 5) = P(3) + P(5)$.

Is it always true that $P(A \text{ or } B) = P(A) + P(B)$?

Suppose we consider the following problem:
"The man I marry will have to be between 25 and 35," decided Jane, "but if he is a millionaire—well, to be more realistic, I suppose an income of £8000 a year or more would do—then I do not really mind how old he is." What is the probability that the first man Jane meets will satisfy her conditions?

If we call A the event that the first man is the right age, and R the event that he is rich enough, then we can find estimates for $P(A)$ and $P(R)$ from a source such as the *Annual Abstract of Statistics*. (We shall assume that we can take E to be the population of British males who have left school.) If we take $P(A) = 0.21$ and $P(R) = 0.02$ (correct to 2 decimal places), is it true that $P(A \text{ or } R) = P(A) + P(R) = 0.23$? Suppose we translate our information to a diagram showing the sets A and R. We cannot assume that the sets A and R do not overlap for $A \cap R = \{$men between 25 and 35 who also earn over £8000 a year$\}$ is unlikely to be an empty set.

The set which will satisfy Jane is $A \cup R$—those who are in the right age group, or rich, or both.

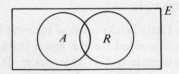

Figure 2.2

$$P(A \cup R) = n(A \cup R)/n(E) \quad \text{by definition,}$$

and clearly $n(A \cup R) = n(A) + n(R) - n(A \cap R)$; for the number of men in $A \cup R$ is found by adding the number in A to the number in R and then subtracting the number in $A \cap R$ as they will have been counted twice.

So, $P(A \cup R) = n(A)/n(E) + n(R)/n(E) - n(A \cap R)/n(E)$
$$= P(A) + P(R) - P(A \cap R)$$

which we can write as
$$P(A \cup R) + P(A \cap R) = P(A) + P(R).$$

So $P(\text{first man is eligible}) = 0.21 + 0.02 - P(\text{first man is right age and rich}).$

As the *Annual Abstract of Statistics* does not give us a breakdown of salary by age-groups, we are unable to find the latter probability immediately. Hence we can only assume now that the probability that the next man Jane meets will satisfy her requirements is rather less than 0·23, the difference being the probability that the man satisfies both of her conditions.

So the general formula, which is the *addition law* is:

$$P(A \text{ or } B) \equiv P(A \cup B) = P(A) + P(B) - P(A \cap B)$$

(This formula is accepted as true for all probabilities.) This reduces to the formula $P(A \cup B) = P(A) + P(B)$ only if $P(A \cap B) = 0$, in other words if the two events A and B cannot both take place at once. In this case the two events are said to be *disjoint* because they give rise to disjoint sets.

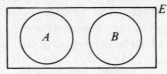

Figure 2.3

We can also see that the result we found in the first chapter for $P(A')$ is a special case of this, for A and A′ are certainly disjoint events and hence $P(A \cup A') = P(A) + P(A')$. But $A \cup A' = E$ since one of A and A′ must happen every time, and $P(E) = n(E)/n(E) = 1$. From this we can easily deduce the result $P(A') = 1 - P(A)$.

To take another example:

The excitement in Little Wickington is running high—one ball of the match to go with 4 runs needed for the home team to win, and Jim, the last man, looking very tense out at the wicket. It is not unknown for Jim to hit a 4, perhaps once in every 20 balls, and he has even managed a 6 on the odd occasion, say with one chance in a hundred. However, as he is not as young as he was, he is not likely to run 4 or more runs, and there is probably an even chance that he will be out anyway. Discounting possibilities of no-balls, wides, etc., what is the probability that (*a*) the home team will win, (*b*) there will be a draw or a tie?

Clearly
$$P(\text{win}) = P(4 \text{ or } 6) = P(4) + P(6) - P(4 \text{ and } 6) = 0·05 + 0·01 - 0 = 0·06.$$

What about the probability of a draw or tie though? If we tackle it directly we would have to consider the probability of drawing and tying separately, and we are not given enough information to do this. We can, however, look at the problem the other way round, and say:

P(draw or tie) $= 1 - P$(neither a draw nor a tie)

$\qquad\qquad\qquad = 1 - P$(win or loss)

$\qquad\qquad\qquad = 1 - P$(win) $- P$(loss) $\qquad$ (since P(win and loss) $= 0$).

P(loss) $= P$(Jim is out) $= 0 \cdot 5$

Hence P(draw or tie) $= 1 - 0 \cdot 06 - 0 \cdot 5 = 0 \cdot 44$.

In this example we notice that P(win)$+ P$(draw or tie)$+ P$(loss) $= 1$. In general if A_1, A_2, ..., A_n are disjoint events so that no pair can happen simultaneously and are also exhaustive, which means that they cover all possible outcomes, then $P(A_1) + P(A_2) + \ldots + P(A_n) = 1$.

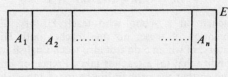

Figure 2.4

EXERCISE 2B

You may find useful the table of log $n!$ on page 247.

1. In Acacia Avenue three housewives have no refrigerator, six own Freezeasys, four Coldomatics and one a Supericer. A market researcher calls at the first three houses. Find the probability that the housewives:
   (i) all own Freezeasys,
   (ii) each own a different make of refrigerator.
   Comment on any assumptions you have made.
2. Calculate the number of different selections of three letters that can be made from the ten letters of the word MANCHESTER. Find, in addition, the probability that any one selection of three letters will contain:
   (i) just one E,
   (ii) at least one E. $\qquad\qquad$ (J.M.B.)
3. If all the sevens are removed from a normal pack of 52 cards, find the probability that a hand of 6 cards drawn from those remaining forms a sequence (i.e. with consecutive values, no account being taken of suit; an ace has value one). $\qquad\qquad$ (J.M.B.)
4. Prove that $\binom{n+1}{r} = \binom{n}{r} + \binom{n}{r-1}$,
   (a) using the factorial definition of $\binom{n}{r}$, (b) using the definition of $\binom{n}{r}$ as "the number of ways of choosing r things out of n".
5. I cannot decide to which of my 7 friends to give Christmas presents. In how many ways could I give k presents, where $k = 1, 2 \ldots 7$? What is the total number of possibilities I have to consider? Does this help to justify the result

$$2^n = \binom{n}{0} + \binom{n}{1} + \ldots \binom{n}{n}?$$

6. In how many ways can 12 people be distributed equally between 3 different cars? What is the probability that 2 specified people are
   (i) in a certain car,
   (ii) together in any car?

7. An encyclopaedia consisting of n ($n \geqslant 4$) similar volumes is kept on a shelf with the volumes in correct numerical order; that is, with volume 1 on the left, volume 2 next, and so on. The volumes are all taken down for cleaning and are replaced on the shelf in random order. Prove that the probabilities of finding exactly n, $(n-1)$, $(n-2)$, $(n-3)$ volumes in their correct positions on the shelf are respectively. ·

$$\frac{1}{n!}; \quad 0; \quad \frac{1}{(n-2)!2}; \frac{1}{(n-3)!3} \qquad \text{(Cam.)}$$

8. Of the 21 teachers in a school who teach History, English or French, or some combination of these, no one teaches both History and French. Eight teach History, of whom 5 do not also teach English; and 7 teach French. Find the probability that (a) a teacher takes French or English, (b) a teacher takes English, given that the probability he takes either History or English is the same as the probability of his taking either French or English.

9. The probability that a girl reads *Woman* is 0·7, and the probability that she reads *Woman* but not *Vogue* is 0·6, and the probability that she reads neither is 0·2. Find the probability that she (a) reads *Vogue*, (b) either reads *Woman* or does not read *Vogue*.

10. If A, B and C are three events, illustrate using Venn diagrams that:
    (i) $P(A) = P(A \cap B') + P(A \cap B)$.
    (ii) $P(A \cap B') = P(A) - P(B) + P(A' \cap B)$
    (iii) $P(A \cup B \cup C) = P(A) + P(B) + P(C) - P(B \cap C) - P(C \cap A) - P(A \cap B) + P(A \cap B \cap C)$.

11. Out of a box containing 6 discs numbered 1 to 6 two discs are drawn without replacement. Find the probability that their sum is (a) 8, (b) less than 10, (c) even, (d) not 7.

12. A restaurant manager works on the assumption that all his customers will have fruit, or cheese and biscuits, or coffee (or any combination), and that the probability of their having fruit is 0·7, fruit and cheese 0·25, cheese and coffee 0·35, and fruit and coffee 0·5. Also the probability of their having fruit or cheese is 0·9, and that of having fruit or coffee is 0·95. Find the probability that they have: (i) cheese or coffee, (ii) coffee only, (iii) all three.

13. Of my ten cups, arranged at random on a shelf, 2 come from a cheap yellow set, 3 are expensive bone china, and the remaining 5 are plastic. While reaching for a plate I knock two of them off the shelf. What is the probability that:

    (i) they are both bone china,
    (ii) they are both of the same kind,
    (iii) they are of different kinds?

14. Find the probability of a Bridge hand (thirteen cards) containing more than 10 trumps.

15. Twenty-five boys and twenty girls set out on a school trip in a 36-seater coach and the remainder in a mini-bus. If the pupils are distributed at random between the vehicles, find the probability that:

   (i) there are no girls in the mini-bus,

  (ii) there are equal numbers of boys and girls in the coach,

 (iii) there are more girls than boys in the mini-bus.

Conditional Probability

3.1. The multiplication law

We have found that in the case of two *disjoint* events A and B we can say that

$$P(\text{A or B}) \equiv P(\text{A} \cup \text{B}) = P(\text{A}) + P(\text{B}).$$

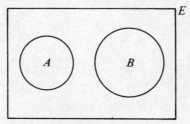

Figure 3.1

This leads us to look at what happens when $P(\text{A} \cap \text{B}) \neq \emptyset$ and the two can therefore occur together. Is there a similar expression for linking $P(\text{A and B}) \equiv P(\text{A} \cap \text{B})$ with $P(\text{A})$ and $P(\text{B})$?

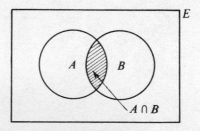

Figure 3.2

George met a dazzling girl at a party but cannot remember the number of the house where she lives. He remembers that it has two digits, which, together with the fact that there are only 39 houses in the road, limits the field. He reckons that the postman will be able to deliver a letter provided the tens digit is correct. What are the chances of his letter getting to the right house? How much worse are his chances if his assumption about the postman is unjustified?

Let A: tens digit is correct,

　　　B: final digit is correct,

and let us assume that George is equally likely to choose any of the possible combinations.

Then A ∩ B: whole address correct. We can use a table to illustrate all the possible cases.

		final digit									
		0	1	2	3	4	5	6	7	8	9
tens digit	1	×	×	×	×	×	×	×	×	×	×
	2	×	×	×	×	×	×	×	×	×	×
	3	×	×	×	×	×	×	×	×	×	×

Figure 3.3

Now $P(A) = \frac{1}{3}$ as we have only three possible figures, each of which is equally likely, and for the same reasons $P(B) = \frac{1}{10}$.

Also with 30 possibilities, $P(A \cap B) = \frac{1}{30}$.

Thus P(finding right house if postman cooperative) $= \frac{1}{3}$ and P(getting exact address) $= \frac{1}{30}$.

So here we have $P(A \cap B) = P(A) . P(B)$.

We can draw a "probability tree" diagram which looks like Fig. 3.4 or Fig. 3.5.

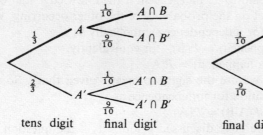

tens digit　　final digit

Figure 3.4

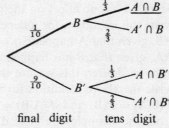

final digit　　tens digit

Figure 3.5

Suppose now that George goes to consult Jim about his problem, to try to narrow the odds against finding the girl. Jim remembers that the figures are both small, either 1, 2 or 3. However, when George suggests 11 as a first try, Jim adds that he is fairly certain that the two digits are different. How much improved are George's chances now?

If we try the table again, we see that
this time our total set of possibilities is
much reduced.

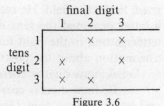

Figure 3.6

Again let A, B be the events defined
before, but under the new circumstances.
We see that $P(A) = \frac{1}{3}$ as there are still
3 equally likely possibilities for the tens
digit, but now $P(B) = \frac{1}{3}$ as the same is true of the final one as well.

$P(A \cap B) = \frac{1}{6}$ since there are six possibilities in all, but $P(A) \cdot P(B) = \frac{1}{3} \cdot \frac{1}{3} = \frac{1}{9}$, so clearly our neat multiplication rule breaks down here. We must therefore analyse the situation carefully to see just where the difference lies between this and the previous case.

One difference is in the appearance of the table; in the first two examples the possibilities could be drawn in a rectangular array, but in the third the reduced table had gaps in the array. This is because in the first example any of the tens digits could be coupled with any of the units; whereas in the second example certain combinations, namely 11, 22 and 33, were not allowed. In other words, in the first case the tens digit is independent of the units digit, but in the second case the tens digits allowed depend on the choice of the units digit, or vice versa.

In general, two events are said to be **dependent** on one another if the knowledge that one has occurred alters the probability of the other occurring. For example, if we know that George has correctly chosen the tens digit, the probability of his second digit being right becomes $\frac{1}{2}$ instead of $\frac{1}{3}$.

Two events are thus said to be **independent** if the knowledge that one has occurred has no effect on the probability of the other occurring. We can write this condition for independence symbolically as

P(B, given that A happened) $= P(B)$,   or equivalently,

P(A, given that B has happened) $= P(A)$.

In order to shorten this we use the sign | to mean "given that". So we can write the above condition for independence as

$P(B|A) = P(B)$ and $P(A|B) = P(A)$.

From the examples we have discussed it looks as if the question of whether or not the multiplication law holds is closely related to whether or not the events concerned are dependent or independent. We will examine this relationship in the next section.

3.2. Conditional probability

The probability of the event B, given that A has happened, $P(B|A)$, is also sometimes called the probability of B **conditional** on A. Let us see if we can derive a formula for this probability by going back to the fundamental definitions of probability which we met in the first chapter.

Suppose we have a set E of equally likely outcomes of an experiment. The event A consists of $n(A)$ of these outcomes and the event $A \cap B$ of $n(A \cap B)$ of them. Then we know that one definition of $P(A)$ is $n(A)/n(E)$.

When we are considering $P(B|A)$ we must assume that A has occurred, so we can now look on A as representing the set of possible outcomes. The event B also occurs if any of the outcomes in $A \cap B$ occur. Hence the total number of possible outcomes conditional on A is $n(A)$, of which $n(A \cap B)$ include the occurrence of event B.

Hence

$$P(B|A) = n(A \cap B)/n(A)$$

$$= P(A \cap B)/P(A)$$

(We have assumed that $P(A) \neq 0$.)

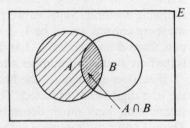

Figure 3.7

Similarly we could have obtained

$$P(A|B) = P(A \cap B)/P(B).$$

We have derived these results using the symmetry definition of probability, but we accept the formula as valid for all probabilities.

Notice that we can rewrite either of these equations as an equation for $P(A \cap B)$ in the form

$$P(A \cap B) = P(B|A)P(A) \text{ or } P(A \cap B) = P(A|B) P(B)$$

and these give us the correct versions of the multiplication law.

Let us see how this works for the final version of George's problem. There we had $P(A) = P(B) = \frac{1}{3}$ and it is easy to see that $P(A|B) = P(B|A) = \frac{1}{2}$. Using either version of the law we thus get $P(A \cap B) = \frac{1}{3} \cdot \frac{1}{2} = \frac{1}{6}$ which we know is correct. We could draw a tree diagram of the situation as in Fig. 3.8 or Fig. 3.9.

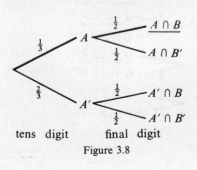

Figure 3.8

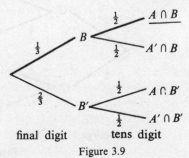

Figure 3.9

We can now state the *multiplication law* as:

$$P(A \text{ and } B) \equiv P(A \cap B) = P(A|B).P(B) = P(B|A).P(A)$$

where $P(A|B)$ is the probability that event A happens given that B has already occurred.

If A and B are *independent*, $P(A|B) = P(A)$ and $P(B|A) = P(B)$ so $P(A \text{ and } B) \equiv P(A \cap B) = P(A).P(B)$.

3.3. Sampling with and without replacement

The situation of George choosing a two-digit number is exactly analogous to drawing numbers out of a hat, or any other random sampling method. Suppose we restrict ourselves to the case when both numbers chosen are to be 1, 2 or 3, then we can imagine that George puts three discs labelled 1, 2 and 3 into a bag. The first one he pulls out is the first digit of his number and the second one the second digit.

If he does not replace the first disc, he has exactly the condition that the first digit is not repeated. Thus what he can pull out of the bag the second time very much depends on what he pulled out the first time. This is known as "sampling without replacement" and is a case where the modified multiplication law will hold.

If he replaces the first disc before he draws the second, however, the number on the first disc will not affect the probability of any of the three numbers being pulled out the second time. This is "sampling with replacement" and here the simple multiplication law will hold.

3.4. A further example

Suppose that George is in the "Red Lion" at closing time two nights out of every three, whereas Jim is there about three times a week and Fred only about once a week. How likely is it that they will meet?

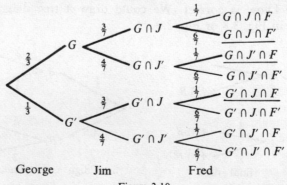

Figure 3.10

If we let G, J and F stand for the events of each being there on a particular night, then we have $P(G) = \frac{2}{3}$, $P(J) = \frac{3}{7}$, $P(F) = \frac{1}{7}$, assuming that none of them have any fixed nights for being or not being there. If we also assume that the three men choose their pub nights independently, then the tree diagram will look like Fig. 3.10.

We can see that $P(G \cap J \cap F) = \frac{2}{3} \times \frac{3}{7} \times \frac{1}{7} = \frac{2}{49}$, so they are there all together roughly once every $3\frac{1}{2}$ weeks. The tree also gives a lot more information; for example, exactly two of them are there on the occasions underlined. So

$$P(\text{exactly two there}) = P((G \cap J \cap F') \cup (G \cap J' \cap F) \cup (G' \cap J \cap F))$$
$$= P(G \cap J \cap F') + P(G \cap J' \cap F) + P(G' \cap J \cap F)$$
$$= \frac{2}{3} \times \frac{3}{7} \times \frac{6}{7} + \frac{2}{3} \times \frac{4}{7} \times \frac{1}{7} + \frac{1}{3} \times \frac{3}{7} \times \frac{1}{7} = \frac{47}{147}$$

(Since the 3 possibilities are disjoint the simple addition law holds.)

So on roughly one night in 3 there are exactly two of them there. We have in fact used the simple multiplication law for independent events extended to three events, i.e.

$$P(A \cap B \cap C) = P(A) \cdot P(B) \cdot P(C)$$

Of course in a real problem it is most unlikely that their visits to the pub would be completely independent; for instance, it might well be that Fred calls for the other two each time he goes to the pub, so that we get a tree like Fig. 3.11 and $P(G \cap J \cap F) = \frac{1}{7}$, since they can only all be

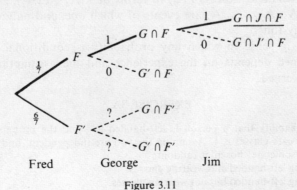

Figure 3.11

there if Fred is, and conversely, if Fred is, they all are, which happens on the one night a week when Fred goes. (Notice that $P(G|F) = P(J|F) = 1$ since, if Fred goes, the other two are certain to do so as well.)

Suppose there is a fourth member of the group, Bill. Fred always tells Bill if he is going to the pub that evening and asks Bill if he would like to be collected. The probability that Bill goes to the "Red Lion" is $\frac{3}{4}$ if

Fred collects him, but only $\frac{1}{3}$ if he has to go on his own. What is the probability that Bill goes to the pub on any particular night?

Again we can draw up a tree (Fig. 3:12).

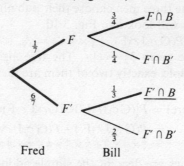

Figure 3.12

What we want is $P(B)$, but we can write this as $P(B \cap F) + P(B \cap F')$ since these events are disjoint. You can see from the tree that this is equal to $(\frac{1}{7} \times \frac{3}{4}) + (\frac{6}{7} \times \frac{1}{3}) = \frac{11}{28}$. So we have $P(B) = \frac{11}{28}$.

In general for two events A and B we have $P(B) = P(B \cap A) + P(B \cap A')$ and we can write this as $P(B) = P(B|A)P(A) + P(B|A')P(A')$, using the multiplication law. This result is often very useful for calculating the probabilities of complicated events. You should be able to see how to extend the result to express $P(B)$ in terms of $P(A_1)$, $P(A_2)$, ..., $P(A_n)$ where A_1, A_2, A_3, ..., A_n are events of which one, and only one, must occur at any time.

There is a sense in which any probability is conditional, since the value assigned depends on the experience and the assumptions of the person concerned.

EXERCISE 3A

1. The probability that a person is left-handed is $\frac{2}{5}$ and the probability that a person wears glasses is $\frac{1}{3}$. Assuming the two are independent, find the probability of someone chosen at random:
   (i) being left-handed and wearing glasses,
   (ii) being left-handed but not wearing glasses,
   (iii) being left-handed or wearing glasses.

2. A random-number table consists of a succession of digits each chosen at random and independently from the set $(0, 1, 2, 3, \ldots, 9)$.
   (a) Two successive digits are taken from such a table; show that the probability that their sum is 9 is $\frac{1}{10}$.
   (b) Four successive digits are taken from the table. Calculate the probability p that the sum of the first two equals the sum of the third and fourth.

(M.E.I.)

3. Two brands of a certain drug are on the market, type X and type Y. One in 300 people taking drug X suffer side-effects and 1 in 900 taking drug Y. At the present time it is estimated that equal numbers of people take each kind of drug. If drug X is taken off the market, prove that the probability of side effects will be halved.

4. The probability that John will go to Cambridge is estimated at $\frac{1}{2}$; the probability that he will go to some other university is $\frac{1}{3}$. The probability that his sister Mary will go to Cambridge is $\frac{1}{4}$. Calculate the probabilities that:
   (a) John and Mary both go to Cambridge,
   (b) John will not go to a university,
   (c) either John or Mary (but not both) will go to Cambridge.   (Cam.)

5. A card is drawn at random from an ordinary pack of cards and E_1, E_2, E_3 are respectively the events that it is a spade, a king, a black card. Prove that E_1 and E_2 are independent, that E_2 and E_3 are independent but that E_1 and E_3 are not independent.   (M.E.I.)

6. A and B are independent events. Show that the following pairs are also independent: A and B′, A′ and B′.

7. A garden has 3 flower beds. The first bed has 50 flowers of which 10 are red, the second bed has 30 flowers of which 10 are red, and the third has 20 flowers of which 10 are red. One of the beds is chosen at random, and a flower is selected at random from that bed and removed from the garden. Find:
   (i) the probability that a red flower will be taken from the first bed,
   (ii) the probability that a red flower will be taken from the garden.
   If a flower is chosen at random from the 100 flowers in the garden, find the probability that the flower will be red.   (M.E.I.)

8. Both the Prime Minister and the Leader of the Opposition are making important speeches tonight. Three newspapers A, B and C differ in their editorial policies, which are independent of each other, so that their editors will allocate the principal front-page headline in tomorrow's issue to the Prime Minister's speech or to the Leader of the Opposition's with the following probabilities. (The headline will not refer to both speeches, and may not refer to either.)

	A	B	C
Prime Minister's speech	1/2	2/5	2/3
Leader of the Opposition's speech	1/5	2/5	1/6

Calculate the probabilities that the headline will be allocated to:
   (i) the Prime Minister's speech in all three papers,
   (ii) the Leader of the Opposition's speech in no paper,
   (iii) one of the speeches in one paper and neither of the speeches in the other two,
   (iv) the Prime Minister's speech in one paper only and the Leader of the Opposition's speech in one paper only.   (M.M.E.)

9. Bag A contains 6 red discs and 4 black discs, bag B contains 2 black discs and 8 red discs.
   (i) One disc is withdrawn from A and left aside, colour unnoticed, and then a second disc is withdrawn from A. Find the probability that it is red.
   (ii) One disc is withdrawn from A and placed in B, then a disc is withdrawn from B. Find the probability that it is red.

   (iii) One of the bags is chosen at random. A disc is withdrawn from it and placed in the second bag, then a disc is withdrawn from the second bag. Find the probability that the second disc is red.

   (iv) One disc is withdrawn from each bag. Find the probability that the two discs are the same colour.

10. At a particular university 70% of the students are reading Arts subjects and 30% are reading Science. Of the Arts students 40% are female, whilst of the Science students 10% are female. Of the men $33\frac{1}{3}$% are in halls of residence, the corresponding figure for women being 50%. Residence in a hall is independent of whether a student studies Arts or Science. Find:

   (i) the probability that a student chosen at random is a male Science student,

   (ii) the probability that a student chosen at random is male,

   (iii) the probability that a male student chosen at random is a Science student,

   (iv) the probability that a Science student chosen at random is female and lives in a hall,

   (v) the probability that a female student, living in a hall, chosen at random is a Science student. (Cam.)

11. In a game between A and B the chances of winning points are as follows:

	A	B
Chance of winning the first point	3/5	2/5
Chance of winning if A won first point	2/3	1/3
Chance of winning if B won first point	1/2	1/2

Find the chance that (i) A wins the first four points, (ii) each wins one of the first two points. (O. & C.)

12. A bag contains 5 white and 3 red balls. Balls are drawn in succession and are not replaced. Show that the chance that the first red ball will appear at the fifth draw is $\frac{3}{56}$. (M.E.I.)

13. A boy and girl play a game of "spotting" approaching red cars when travelling along a road on which 25% of the cars are red. If a car is red, the first one to "claim" it gains one point (a claim is always made by either the boy or the girl for a red car, but never by both simultaneously). If a car is not red, it is found that on half the occasions either the boy or the girl (but not both) makes a claim, for which there is a penalty of two points. The first to be five or more points ahead is the winner.

   The events "a claim by the boy" and "the car is red" are independent, and the probability that he makes a claim on a car is $\frac{2}{5}$.

   (i) Draw a tree diagram in which the first two branches lead to the events "red" and "not red", and mark on the secondary branches the probabilities of claims by the children.

   (ii) If it is known that a claim has been made for a car which is not red, what is the probability it was made by the girl?

   (iii) If the score is 8–5 in favour of the girl, show that the probability that she will have won by the time two more cars pave passed, is 0·48. (S.M.P.)

14. A game is played by two players A and B who have alternate turns. When it is A's turn there is a probability of $\frac{1}{3}$ that he will win that turn; otherwise it is a draw. When it is B's turn there is a probability of $\frac{1}{4}$ that he will win; otherwise it is a draw. The game continues until either A or B wins.

   (i) If A starts, show that the probability of B winning at his first turn is $\frac{1}{6}$.

   (ii) If B starts, find the probability that he wins at his next turn.

(iii) If A starts, find the probability that he wins at either his first or second turn.

(iv) If A and B toss an unbiased coin to decide who starts, find the probability that after two turns each neither has yet won.   (O. & C.)

15. If A and B are independent and A is a subset of B, show that $P(A) = 0$ or $P(B) = 1$.

16. Two cards are drawn, at random and without replacement, from a pack of 20 different cards which are numbered 1, 2, ..., 20. Calculate the probabilities $P(A)$, $P(B)$, $P(A|B)$, $P(B|A)$ in each of the following cases:

(i) A is the event "first card is 5".
   B is the event "second card is even".

(ii) A is the event "first card is less than or equal to 5".
   B is the event "second card is greater than or equal to 5".   (Cam.)

17. In a game for two players a turn consists of throwing a dice either once or twice, once if the score obtained is less than 6, twice if the score at the first throw is 6. The score for the turn is in the first case the score of the single throw, and in the second case the total score of the two throws. Obtain the probabilities of a player:

(a) scoring more than 9 in a single turn,

(b) scoring a total of more than 20 in two succeeding turns,

(c) obtaining equal scores in two succeeding turns.   (Cam.)

18. Three cards are drawn from a pack of 52 cards and, when they have been replaced and the pack shuffled, a second set of three cards is drawn. Find the chance that the six cards drawn should include at least one ace.   (M.E.I.)

19. In an experiment for testing the learning ability of rats, a rat is placed at the base of a T-shaped maze and some food at the end of the right arm of the T. On each trial the rat runs along from the base of the maze and must turn to the right or the left. The probability that the rat turns left at the first trial is assumed to be 0·5. Every time the rat turns left the probability that it turns left on the next trial is decreased by 20%; every time it turns right the corresponding probability is decreased by 40%. Draw a tree diagram and find the probability that the rat turns left (a) at the second trial, (b) at the third trial.

20. There are 5 runners in the first race, 6 runners in the second, and 4 in the third. If I bet on each race, find the probability that I back at least one winner, assuming that each horse is equally likely to win its race.

3.5. More conditional probability: Bayes' theorem

"I seem to be a very bad judge of the weather. I drag my umbrella with me three-quarters of the time and yet, although it only rains on a quarter of the days when I take it, it rains on two-thirds of the days when I do not take it. So on a rainy day I always seem to get wet." How true is this?

Suppose we let the events be R: it rains,
                            U: I take my umbrella.
Then $P(U) = \frac{3}{4}$. Also, as it rains on one-quarter of the days when I have

the umbrella and on two-thirds of the days when I do not, $P(R|U) = \frac{1}{4}$ and $P(R|U') = \frac{2}{3}$.

We can use this information to draw up a diagram, which makes clearer the steps we will have to follow.

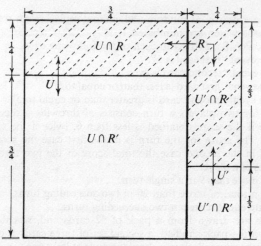

Figure 3.13

We want to know the probability that I get wet on a day when it rains, which is $P(U'|R)$. We can see that $P(U'|R) = P(U'\cap R)/P(R)$.

However the set R is divided into two parts, $U \cap R$ and $U' \cap R$. Using the result of §3.4, we can say that

$$P(R) = P(R|U)P(U) + P(R|U')P(U') = \frac{1}{4} \times \frac{3}{4} + \frac{2}{3} \times \frac{1}{4} = \frac{17}{48}.$$

Hence $P(U'|R) = P(U'\cap R)/P(R) = \dfrac{\frac{2}{3}\times\frac{1}{4}}{\frac{17}{48}} = \frac{8}{17}.$

So, although I carry my umbrella three-quarters of the time, I get wet on 8 out of every 17 rainy days!

We can put some of the algebraic expressions we used together to give the formula:

$$P(U'|R) = \frac{P(R|U')P(U')}{P(R)} = \frac{P(R|U')P(U')}{P(R|U')P(U') + P(R|U)P(U)}$$

or more generally $P(A|B) = \dfrac{P(B|A)P(A)}{P(B|A)P(A) + P(B|A')P(A')}$

which is known as *Bayes' Theorem*.

$P(A)$ is often referred to as the *prior* probability of A, whereas $P(A|B)$ is called the *posterior* probability of A, given B.

Again, we can easily extend Bayes' theorem to the case where, instead of just A and A′, we have a set of possible events A_1, A_2, ... A_n, which are both disjoint and exhaustive.

3.6. Markov chains

"Mathematics examinations are all a matter of luck really," said George. "If I happen to get the first question out, then it is more likely that I will finish the second; if I get the second right, it is more likely that I will finish the third, and so on. But if I cannot do the first question, I get in a panic and then I probably find I cannot do the second or the third or the fourth."

Suppose we examine George's situation in the examination. Let us say that the probability of getting a question right if the previous one was right is $\frac{3}{5}$, but the probability of getting it right if the previous one was wrong is only $\frac{1}{5}$. (We are assuming that he knows when he has got a question right and that each question is only directly influenced by the one before it.) If the probability of his getting the first one right is $\frac{1}{2}$, what is the probability of getting the second right? What about the third?

Let A_1, A_2 and A_3 be the events that he gets the first, second and third questions right respectively. Then we know $P(A_1) = \frac{1}{2}$ and $P(A_2|A_1)$ $= P(A_3|A_2) = \frac{3}{5}$. Also $P(A_2|A_1') = P(A_3|A_2') = \frac{1}{5}$.

If we draw a tree we find:

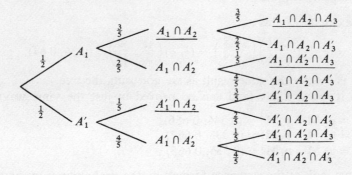

Figure 3.14

We have used the fact that $P(A_2'|A_1) = 1 - P(A_2|A_1) = \frac{2}{5}$.
We can see from the tree that
$$P(A_2) = \frac{1}{2} \cdot \frac{3}{5} + \frac{1}{2} \cdot \frac{1}{5} = \frac{4}{10} = 0 \cdot 4$$
$$P(A_3) = \frac{1}{2} \cdot \frac{3}{5} \cdot \frac{3}{5} + \frac{1}{2} \cdot \frac{2}{5} \cdot \frac{1}{5} + \frac{1}{2} \cdot \frac{1}{5} \cdot \frac{3}{5} + \frac{1}{2} \cdot \frac{4}{5} \cdot \frac{1}{5} = \frac{18}{50} = 0 \cdot 36.$$

It appears to be decreasingly likely that George gets the questions right as $P(A_1) = 0.5$, $P(A_2) = 0.4$, $P(A_3) = 0.36$. Can we work out what would happen at the 10th question?

Doing it by the tree method might appear very complicated, but if we look carefully we can see a pattern emerging.

$$P(A_1) = \tfrac{1}{2} \qquad\qquad\qquad P(A_1') = \tfrac{1}{2}$$

$$P(A_2) = \tfrac{1}{2}.\tfrac{3}{5} + \tfrac{1}{2}.\tfrac{1}{5} \qquad\qquad P(A_2') = \tfrac{1}{2}.\tfrac{2}{5} + \tfrac{1}{2}.\tfrac{4}{5}$$

$$= (\tfrac{1}{2}, \tfrac{1}{2})\begin{pmatrix} \tfrac{3}{5} \\ \tfrac{1}{5} \end{pmatrix} \qquad\qquad = (\tfrac{1}{2}, \tfrac{1}{2})\begin{pmatrix} \tfrac{2}{5} \\ \tfrac{4}{5} \end{pmatrix}$$

Putting this together into matrix notation we arrive at

$$(P(A_1), P(A_1')) = (\tfrac{1}{2}, \tfrac{1}{2})$$

$$(P(A_2), P(A_2')) = (\tfrac{1}{2}, \tfrac{1}{2})\begin{pmatrix} \tfrac{3}{5} & \tfrac{2}{5} \\ \tfrac{1}{5} & \tfrac{4}{5} \end{pmatrix}$$

Notice that the matrix $\begin{pmatrix} \tfrac{3}{5} & \tfrac{2}{5} \\ \tfrac{1}{5} & \tfrac{4}{5} \end{pmatrix}$ is made up from $\begin{pmatrix} P(A_2|A_1) & P(A_2'|A_1) \\ P(A_2|A_1') & P(A_2'|A_1') \end{pmatrix}$

If we call this matrix $\mathbf{M}$ and the vectors $(P(A_1), P(A_1'))$, $(P(A_2), P(A_2'))$, etc., $\mathbf{p}_1$, $\mathbf{p}_2$ ... we have $\mathbf{p}_2 = \mathbf{p}_1\mathbf{M}$.

But $P(A_3) = P(A_3|A_2)P(A_2) + P(A_3|A_2')P(A_2') = \mathbf{p}_2\begin{pmatrix} \tfrac{3}{5} \\ \tfrac{1}{5} \end{pmatrix}$

and $P(A_3') = \mathbf{p}_2\begin{pmatrix} \tfrac{2}{5} \\ \tfrac{4}{5} \end{pmatrix}$.

So $\mathbf{p}_3 = \mathbf{p}_2\mathbf{M} = \mathbf{p}_1\mathbf{M}^2$.

We can check this as

$$\mathbf{p}_3 = (\tfrac{1}{2}, \tfrac{1}{2})\begin{pmatrix} \tfrac{3}{5} & \tfrac{2}{5} \\ \tfrac{1}{5} & \tfrac{4}{5} \end{pmatrix}\begin{pmatrix} \tfrac{3}{5} & \tfrac{2}{5} \\ \tfrac{1}{5} & \tfrac{4}{5} \end{pmatrix} = (\tfrac{1}{2}, \tfrac{1}{2})\begin{pmatrix} \tfrac{11}{25} & \tfrac{14}{25} \\ \tfrac{7}{25} & \tfrac{18}{25} \end{pmatrix} = (0.36, 0.64)$$

which is exactly the same result as we got using the tree.

So it seems likely (and can be proved in just the same way) that

$$\mathbf{p}_4 = \mathbf{p}_3\mathbf{M} = \mathbf{p}_1\mathbf{M}^3 = (0.344, 0.656);$$
$$\mathbf{p}_5 = \mathbf{p}_4\mathbf{M} = \mathbf{p}_1\mathbf{M}^4 = (0.338, 0.662);$$
$$\mathbf{p}_6 = \mathbf{p}_5\mathbf{M} = \mathbf{p}_1\mathbf{M}^5 = (0.335, 0.665) \ldots.$$

If we look at the sequence formed by the probabilities we have 0.5, 0.4, 0.36, 0.344, 0.338, 0.335 .... It looks as though the sequence is decreasing, and as the difference between the terms is also decreasing, it appears as if the sequence is tending to a limit. In fact $P(A_{10}) = 0.333$ and $P(A_{100}) = 0.333$ (both correct to 3 decimal places). So it seems that the more questions George attempts, the nearer his probability of getting one right gets to $0.3333 \ldots$ or $\tfrac{1}{3}$.

Suppose the probability of his getting a question right were exactly $\frac{1}{3}$. Then the probability vector referring to the next question is

$$\mathbf{p} = (\tfrac{1}{3}, \tfrac{2}{3})\begin{pmatrix} \tfrac{3}{5} & \tfrac{2}{5} \\ \tfrac{1}{5} & \tfrac{4}{5} \end{pmatrix} = (\tfrac{1}{3}, \tfrac{2}{3}).$$

In other words the probability of finishing the next question, and the one after that and so on are all the same. $\mathbf{p}$ is called the *equilibrium* or *stable* state and is the one to which $\mathbf{p}_1, \mathbf{p}_2 \ldots$ tend.

We can find the equilibrium value of $\mathbf{p}$ directly without working out the sequence by noting that at equilibrium $\mathbf{p} = \mathbf{p}\mathbf{M}$; for if $\mathbf{p} = (p, q)$, where $p + q = 1$,

then $(p, q) = (p, q)\begin{pmatrix} \tfrac{3}{5} & \tfrac{2}{5} \\ \tfrac{1}{5} & \tfrac{4}{5} \end{pmatrix} = (\tfrac{3}{5}p + \tfrac{1}{5}q, \tfrac{2}{5}p + \tfrac{4}{5}q).$

So $p = \tfrac{3}{5}p + \tfrac{1}{5}q$, i.e. $2p = q$ and therefore $p = \tfrac{1}{3}$, $q = \tfrac{2}{3}$.

Notice that in working out the equilibrium state we have only used the matrix $\mathbf{M}$, so that the result is independent of the probability of getting the first question right. We can check this using a different value for $P(A_1)$; the sequence $P(A_1)$, $P(A_2) \ldots$ will still tend to $\tfrac{1}{3}$.

This sort of situation where the probability of the nth event (or an event at the nth instant of time) depends only on the result of the $(n-1)$th event (or the event at the $(n-1)$th instant), and the dependence relation is the same for each successive pair of events, is called a (homogeneous) **Markov process** or **Markov chain**.

The probability matrix $\mathbf{M}$ is called the *transition matrix*, as it describes the probabilities of transitions from one "state" to the next. In this example the two possible states after $n-1$ events are "nth question correct", and "nth question incorrect". The matrix need not necessarily be 2×2 but could be 3×3 or 4×4 if there were 3 or 4 possible states at each step.

The Markov chain process has a number of important applications, particularly in the theory of queues and in biology where, for example, the propagation of different genetic types can be described in this way.

EXERCISE 3B

1. I estimate that it rains for at least part of one third of the days of the year. If it rains, the probability that my barometer will have indicated rain the day before is $0 \cdot 7$. If it is a fine day, the probability that the barometer will have indicated rain is $0 \cdot 1$. My barometer is showing fair today, what is the probability that it will rain tomorrow?

2. An engineer is responsible for servicing three small computer installations A, B, C. The probabilities that A, B, C will break down on any day are $\frac{1}{40}$, $\frac{1}{50}$, $\frac{1}{100}$ respectively. The probabilities that the three computers will not operate after 7 pm are $\frac{5}{9}$, $\frac{2}{3}$, $\frac{3}{4}$ respectively. On one day the engineer receives a message that one of the computer installations has broken down at 8 pm but the message does not specify which computer. Calculate the probabilities that the breakdown is at A, B, C, and so decide the first installation which should be checked by the engineer. (M.E.I.)

3. John washes up after lunch three times a week, and his sister Mary washes up the other four times. John's days are chosen at random each week. The probability that he will break one or more dishes during a washing is 0·1 and the probability that Mary will is 0·05. One day after lunch Dad, hearing a dish crash, said: "Apparently this is John's day for doing the washing up." What is the probability that he was right? (S.M.P.)

4. A company advertises a professional vacancy in three national daily newspapers A, B, C which have readerships in the proportions 2, 3, 1 respectively. From a survey of the occupations of the readers of these papers it is thought that the probabilities of an individual reader replying are 0·002, 0·001, 0·005 respectively. (i) If the company receives one reply, what are the probabilities that the applicant is a reader of papers A, B, C respectively? (ii) If two replies are received, what is the probability that both applicants are readers of paper A? (You may assume that each reader sees only one paper.) (M.E.I.)

5. An electronic audio apparatus is liable to two faults T and R (respectively failure of a transistor and failure of a resistor), which occur independently with respective probabilities 0·2 and 0·1. Verify that the prior probabilities of the four states N (fault-free), T, R, and TR (i.e. T and R simultaneously) are as given in the table below. The table also gives the probabilities, conditional on its state, of the apparatus emitting a hum H, a whine W, or a squeal S. Draw up a table giving the posterior probabilities of the four states conditional on H, W or S.

State	Prior probability	Conditional probability			Total
		H	W	S	
N	0·72	1·0	0	0	1·0
T	0·18	0·5	0·4	0·1	1·0
R	0·08	0·7	0·2	0·1	1·0
TR	0·02	0·2	0·3	0·5	1·0

(Cam.)

6. A bag A contains 5 similar balls, of which 3 are red and 2 black, and a bag B contains 4 red and 5 black balls. A ball is drawn at random from A and placed in B; subsequently a ball is drawn at random from B and placed in A. What is the chance that if a ball is now drawn from A it will be red? If this ball is in fact red, what is the probability that the first two balls drawn were also red? (M.E.I.)

7. A certain rare disease occurs in 0·01% of the population. A test to discover the disease gives a positive reaction for 90% of the people suffering from the disease, but also for 2% of people not having it. If a randomly selected person has a positive test, what is the probability that he has the disease?

8. Little Wickington football team have two goal-keepers, the regular player and a reserve, but they are both injured. They estimate that the probability that they will win their next match is $\frac{5}{8}$ if the regular goal-keeper plays, $\frac{1}{2}$ with the reserve, and $\frac{3}{8}$ with anyone else. If the probability that the regular will be fit in time to play is $\frac{1}{2}$, and the probability that the reserve will be fit to play is $\frac{3}{4}$, find the probability that they will win the match. If you knew that they had won the match, what would you give as the probabilities of the regular and reserve goal-keeper having played?

9. If I am late for school I make a greater effort to arrive on time the next school day. If I arrive on time I am liable to be less careful the next day. Consequently, if I am late one day, the probability that I am on time the next day is $\frac{3}{4}$. If I am on time one day the probability that I am late the next day is $\frac{1}{2}$. I am on time on Monday. Calculate the probability that in the same week I shall be on time (i) on the Wednesday, (ii) on the Friday. Show that in the long run I shall probably be on time $1\frac{1}{2}$ times as often as I am late. (M.M.E.)

10. The probability of a team winning a match is $0\cdot6$ and of drawing $0\cdot3$ if the previous match was won. If the previous match was drawn, the corresponding probabilities are $0\cdot2$ and $0\cdot6$, and if it was lost $0\cdot2$ and $0\cdot4$. Find the transition matrix, and hence the probabilities of winning or drawing any particular match in the distant future if the probabilities remain the same. (S.M.P.)

REVISION EXAMPLES I

1. A man tosses a coin repeatedly and scores 1 for each head and 2 for each tail. If p_n is the probability that his score will ever be n, prove that:

$$2p_n = p_{n-1} + p_{n-2}.$$

Prove that $p_n = A + B(-\frac{1}{2})^n$ and evaluate A and B. (M.E.I.)

2. 100 students sit an examination in English, Geography and Mathematics. The probability of a pass in three subjects is $0\cdot29$ and the probability of complete failure is $0\cdot10$. The unconditional probability of a pass in Mathematics is $0\cdot54$. The probability of passes in both English and Geography is $0\cdot33$, and in both Geography and Mathematics is $0\cdot41$ irrespective of the result in the third subject in each case. On average, 14% pass in Geography only and 19% pass in Mathematics without passing in English. Calculate the number of students expected to pass the English examination. (M.E.I.)

3. Seven pieces of luggage are put at random in a row on a truck. What is the chance that two particular pieces will be put next to each other? Calling this probability p, find an expression, in terms of p, for the probability that out of three occasions the pieces should be next to each other exactly twice.
(St. Dunstan's)

4. A box contains four white balls and one black ball, and a second box contains nine white balls and one black ball. An experiment is performed in which a ball, selected at random from the first box, is transferred to the second and at the same time a ball, selected at random from the second box, is transferred to the first. The experiment is termed a success if it results in one of the boxes containing both black balls. Prove that the probability of a success is $0\cdot26$.

The experiment is performed four times. Calculate the separate probabilities of obtaining no successes and one success in the series of experiments, giving each answer correct to two decimal places. Show that the probability of obtaining less than two successes is about 0·72. (M.M.E.)

5. A restaurateur has four times as many male customers as female; 40% of men and 70% of women take the set lunch, the remainder choosing from among the optional items of the menu. Of men choosing the set lunch, 10% drink wine, 50% beer and the remainder a soft drink, whilst of those choosing the optional items the corresponding proportions are 60% and 30%. Among the women customers the corresponding proportions are 30% and 10% of those taking the set lunch and 40% and 20% of those choosing optional items.
   (i) What proportion of customers take the set lunch?
   (ii) What proportion of these drink wine?
   (iii) What proportion of women customers drink beer?
   (iv) What proportion of wine-drinking customers are men? (Cam.)

6. Assuming a population in which equal numbers of births occur in each of the twelve months of the year, calculate the probabilities that, of four persons taken at random (i) no two will be found to have birthdays in the same month, (ii) exactly two will have birthdays in the same month and the others in different months. (M.E.I.)

7. A certain number r of persons sit down in a random arrangement on chairs labelled with their names. If u_r denotes the number of ways that all r chairs can be filled wrongly, write down the values of u_2 and u_3, and calculate u_4. Show that, when $r = 5$, the chance that more than one person is in his right chair slightly exceeds 0·25. (Cam.)

8. A and B are two events and B′ is the complementary event to B (i.e. not B). Show that $P(A)$ lies between $P(A|B)$ and $P(A|B')$. (M.E.I.)

9. A farmer keeps two breeds (A, B) of chicken; 70% of the egg production is from birds of breed A: Of the eggs laid by the A hens, 30% are large, 50% standard, and the remainder small; for the B hens the corresponding proportions are 40%, 30%, and 30%. Egg colour (brown or white) is manifested independently of size in each breed; 30% of A eggs and 40% of B eggs are brown. Find:
   (i) the probability that an egg laid by an A hen is large and brown,
   (ii) the probability that an egg is large and brown,
   (iii) the probability that a brown egg is large,
   (iv) which size contains the largest proportion of brown eggs,
   (v) whether colour and size are manifested independently in the total egg production. (Cam.)

10. I have a coin which I assume is equally likely to be fair ($p = \frac{1}{2}$) or biased ($p = \frac{2}{3}$) where p is the probability of a head in each case. I toss the coin three times and obtain 2 heads. Find the posterior probability that the coin is fair. If I toss the coin another three times and this time obtain one head, find the corresponding probability after both experiments.

11. X, Y and $(n-2)$ other players play a game in which each player has an equal chance of winning each round. The game is won by winning 8 consecutive rounds of the game. Let
$$q_k = P(\text{X wins}|\text{Y has won the last } k \text{ times}),$$
$$r_k = P(\text{X wins}|\text{X has won the last } k \text{ times}), \text{ for } k = 1, 2, \ldots, 7.$$

Show that $(n-1)q_k + r_k = 1$, $r_k = (1/n)\,r_{k+1} + (1-1/n)\,q_1$, for $k = 1, 2, \ldots, 6$, and find the corresponding results for $k = 7$. Hence show that the chance that X will win the game is $(n^7 - 1)/(n^8 - 1)$ if Y has just won one round but did not win the round before that.   (M.E.I.)

12. Three events A, B, C are said to be *independent* if:
    (i) $P(A \cap B) = P(A)P(B)$,   (ii) $P(A \cap C) = P(A)P(C)$,
    (iii) $P(B \cap C) = P(B)P(C)$,   (iv) $P(A \cap B \cap C) = P(A)P(B)P(C)$.
    If only the first three conditions hold, the events are said to be *pairwise independent*. Give examples of:
    (a) 3 events which are pairwise independent but not independent,
    (b) 3 events for which condition (iv) holds but which are not independent.
    (*Hint*: for case (a) consider choosing at random an integer from the set 1 to 9; for case (b) from the set 1 to 12.)

13. Three bags A, B, C contain respectively 3 white and 2 red balls, 4 white and 4 red balls, 5 white and 2 red balls. A ball is drawn unseen from A and placed in B; then a ball is drawn from B and placed in C. Find the chance that if a ball is now drawn from C it will be red.   (M.E.I.)

14. In lawn tennis a set is won by the first player to win six games, except that if the score reaches 5–5 the set is won by the first player to lead by 2 games. Two players have chances respectively p and q of winning in any game ($p + q = 1$); games may be treated as independent. Find the chance that a set lasts exactly $2n + 2$ games ($n \geqslant 5$).   (Cam.)

15. In an examination the respective probabilities of three candidates solving a certain problem are $\frac{4}{5}$, $\frac{3}{4}$ and $\frac{2}{3}$. Calculate the probability that the examiner will receive from these candidates (i) one, and only one correct solution, (ii) not more than one correct solution, (iii) at least one correct solution.
                                    (J.M.B.)

16. Half the population of the city of Ekron are Philistines and the other half are Canaanites. A Philistine never tells the truth, but a Canaanite speaks truthfully with a probability $\frac{2}{5}$ and falsely with a probability $\frac{3}{5}$. What is the probability that a citizen encountered at random will give a correct answer to a question? Tabulate the probabilities that 0, 1, 2 or 3 men out of a sample of 3 citizens taken at random from Ekron will affirm a proposition:
    (a) when it is true,   (b) when it is false.
    The proposition has a prior probability $\frac{3}{4}$ of being true. What is the posterior probability of it being true conditional on it being affirmed by only one out of the three?   (Cam.)

17. In a game of Bridge, each of four players is dealt 13 cards after the pack has been shuffled. One of the players has four diamond cards, and he can see four more diamonds in the exposed dummy hand which has been put down on the table by one of the players. What is the chance that the five remaining unseen diamond cards are split three in one hand and two in the other? If, in the first three rounds of play, each player has used exactly one diamond and two other cards, what is the chance that one of the players has no diamond cards remaining in his hand?   (M.E.I.)

18. The blood of people is either Rhesus-positive or Rhesus-negative. Which it is, is determined by two genes R and r. Each person has two genes, one inherited from each parent; only those with two r genes are Rhesus-negative. There is a medical problem if a Rhesus-negative mother gives birth to a Rhesus-

positive child. Assuming that a parent is equally likely to pass either of its genes to a child and that the proportions of English people with the three gene combinations are RR 36%, Rr 48%, rr 16%, find the probability of a Rhesus-negative mother giving birth to a Rhesus-positive child given:

(a) the mother is known to be Rhesus-negative and the father Rhesus-positive,

(b) the mother is known to be Rhesus-negative but nothing is known about the father's blood-group,

(c) nothing is known about the blood-group of either.

Find also the probability in case (a) if you know that the mother already has one Rhesus-positive child.

19. Two players in a tennis tournament with 16 competitors are placed in different halves of the draw. If the probability that the first player A beats the second player B is r, and the probability that A beats any other player is p, while the probability that B beats any other player is q, find the probability that A wins the tournament. Find also the probability that A wins the tournament if the players are placed randomly in the draw.

20. The probability that I have a bath on any day is governed only by whether or not I have had baths in the previous two days. The probability is 0·4 if I have had baths on both previous days, 0·5 if I bathed yesterday but not the day before that, 0·6 if I bathed two days ago but not yesterday and 0·8 if I bathed on neither day. Show that the problem can be considered as a four-state Markov chain and find its transition matrix. Hence or otherwise find the probability that I have a bath on (i) the Thursday, (ii) the Friday of a week in which I bathed on both Monday and Tuesday. Find also the proportion of days in which I have a bath in the distant future if the probabilities remain the same.

INVESTIGATIONS II

The next three chapters deal with the analysis of experimental data. Later ones go on to discuss the fitting of theoretical distributions to the data in order to use them as a basis for statistical decision. In order to make this more meaningful, it is suggested that at least one of the following investigations is carried out, or any other collection of data (e.g. scientific, economic, etc.) which results in a frequency table of some kind of measurement is considered. These data can then be used for analysis in order to apply the ideas of the following chapters in a real situation. The investigation should involve (a) planning the experiment—how to collect the data, how much to collect, etc., (b) displaying the results in clear tabular and diagram matic form, (c) analysing the data with the help of Chapters 4–6, (d) writing a critical account of the procedure and drawing any conclusions which are possible. Investigate:

1. Queues at a (a) road junction, (b) supermarket till, (c) post office counter, (d) garage, (e) bus stop. This may include length of queue, intervals between arrivals, length of serving time, or length of time any individual has to wait.

2. Error involved in (a) guessing the length of a line, height of a building, weight of a cake, etc., (b) measuring the length of a room, field or street or the weight of a piece of lead shot, grain of rice, etc., (c) aiming for the centre of a dartboard, (d) dropping rice grains so as to land on a particular line.

3. The number of (a) children in a family, (b) houses in a street, (c) the house in which a person lives, (d) goals in a football match, runs in a cricket innings, etc., (e) draws in the first division, (f) words in a sentence or letters in a word (try comparing different authors), (g) free spaces in a small car park, (h) matches (or drawing pins, etc.) in a new box, (i) heads appearing when 8 coins are tossed, (j) throws before a 6 first appears on a die or a head is obtained with a penny, (k) people absent from a class, school or place of work, (l) diamonds in a hand of 13 cards, (m) white beads in a sample of 10 from a mixed bag of black and white, (n) figures people can remember of a long telephone number, (o) drawing pins, pegs, forks, etc., which fall in a particular way from a sample of 10, (p) woodlice preferring the dark (or damp) half of a box out of a sample of 10, (q) paces a drunkard is from his starting point if he goes backwards or forwards at random.

4. The (a) ages of cars, (b) girths of trees, (c) marks obtained in an examination, (d) lengths of middle fingers (or heights, weights, shoe sizes, etc.), (e) speeds of cars on a particular stretch of road, (f) amounts on bills in a supermarket, (g) weights of books, (h) times people take to fulfil a task, e.g. a cross-country run, a row of knitting, a mathematical calculation, (i) distances between the two ends of a metre of string when it is dropped in a random way on a flat surface.

5. The (a) last figure of a telephone number, (b) score obtained with 2 or more dice, (c) difference between the numbers on two halves of a domino, (d) frequencies of random numbers in the table on page 244, (e) gaps between cars on a fairly busy road, (f) number of occurrences of a particular word on a page.

Frequency Distributions
Modes and Means

4.1. Frequency charts: golf scores

Traditionally statistics is thought of as the subject concerned with analysing data obtained from experiments, and it is this aspect we shall be concerned with in the next three chapters. At the beginning of this chapter are a number of investigations which will provide data to be analysed. It is well worth carrying out some of these, as it is always more interesting to handle data you have personally collected. In this chapter and the next two we simply look at three specimen cases to see how we might proceed.

During a rather bleak week at a British seaside resort Jim found himself becoming a miniature golf enthusiast. Suppose we want to analyse one particular morning's performance for which his score card, after 2 rounds, read:

$$4 \quad 3 \quad 4 \quad 4 \quad 3 \quad 5 \quad 4 \quad 4 \quad 3 \quad 3 \quad 3 \quad 4 \quad 5 \quad 3 \quad 4 \quad 3 \quad 6 \quad 6$$
$$3 \quad 4 \quad 4 \quad 3 \quad 4 \quad 2 \quad 7 \quad 6 \quad 4 \quad 3 \quad 5 \quad 6 \quad 5 \quad 4 \quad 3 \quad 6 \quad 3 \quad 7$$

In order to extract a clearer picture, it is probably more meaningful to list the frequency with which each score occurs in a **frequency distribution** or **frequency table**:

score (x)	frequency (f)
1	0
2	1
3	12
4	12
5	4
6	5
7	2
8 and over	0
	36

In the sequence of investigations in Chapter 1 we were interested in finding the probability of an event by counting the number of times it did or did not happen. Usually however, as here, we want to look at the overall pattern of the frequency with which each of the possible outcomes occurs.

We can make the pattern even easier to assimilate by drawing a picture of it. If you look through newspapers and statistical reports you will find the two most common methods are the pie chart and the frequency chart.

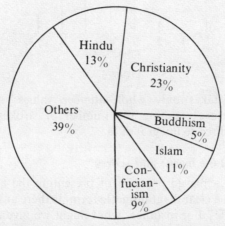

Figure 4.1—Pie chart illustrating the relative membership of world religions ('others' includes atheists, etc.).

A typical pie chart is shown above; the topic may be a far cry from miniature golf, but would the method be equally suitable? There would be certain disadvantages: we would probably lose the significance of the exact frequencies as only the relative proportions are depicted, and we would in addition lose any reference to scores such as 1 and 8 which, although possible, did not occur in this particular case. Also, and perhaps most important, the pattern of increasing frequencies up to 3 and decreasing from 4 onwards would not be easy to pick out.

For such reasons the frequency chart is more useful statistically.

Figure 4.2 shows the pattern much more clearly. Sometimes the lines are widened into columns on such a chart. We have preferred not to do so in this case as the score can only take values which are whole numbers and cannot take any values between, like 2·5, which might be suggested by wider columns. Such cases where there are gaps between each of the values taken by the variable are called **discrete**. It often happens that a

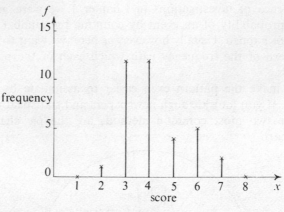

Figure 4.2

discrete variable takes only whole-number values; e.g. a counting process (here we are counting the number of strokes at each hole) always produces discrete integral data.

4.2. The mode and the mean: golf scores

We have now achieved one aim of presenting the analysis of Jim's golfing prowess in a clear diagrammatic form. Is there any other information we might note? For instance, what might we give as a measure of his achievement if we wished to compare it with previous rounds?

We might take as "typical" the score which occurs most frequently, which is called the **mode** (or modal value). The word is related to the French *à la mode* which means "fashionable", as we are choosing the "most fashionable" value when we take the one which has the highest frequency. This has an obvious drawback in this particular case as we are left with a choice between taking 3 or 4 as our mode; when this happens the distribution is sometimes described as *bimodal*. To compromise and say that the mode is $3\frac{1}{2}$ would be meaningless, as we are considering a typical value for the number of strokes.

The other obvious measure that might be of interest is the "average" score, which we shall in fact refer to as the *arithmetic mean* or simply as the **mean**. We can obtain this by adding up all 36 of the scores and dividing by 36. As 2 occurs once, 3 occurs twelve times, and so on, the total sum of scores $= 1 \times 2 + 12 \times 3 + 12 \times 4 + 4 \times 5 + 5 \times 6 + 2 \times 7 = 150$. Hence the mean score, which we shall call $\bar{x}$ (pronounced x-bar), will be given by:

$$\bar{x} = (1 \times 2 + 12 \times 3 + 12 \times 4 + 4 \times 5 + 5 \times 6 + 2 \times 7)/36 = 150/36 \simeq 4 \cdot 17.$$

So the mean score is 4·17. Notice that the mean, unlike the mode, is not necessarily equal to one of the possible scores.

Suppose we look at how we worked out $\bar{x}$. In order to obtain the total of the scores, we in fact simply multiplied each possible score x by its frequency f, which is the value opposite it in the frequency table. We then added up the resulting values and divided by the total number of scores, n say. If we want to write this out as a formula, we can write it as:

$$\bar{x} = (f_1 x_1 + f_2 x_2 + f_3 x_3 + \ldots)/n$$

where f_1 is the frequency of value x_1 and so on.

In order to save writing all this out we use the sign Σ (sigma, the Greek capital S) to stand for "sum of": for instance Σf means "add up all the values of f". So in this particular case $\Sigma f = 1 + 12 + 12 + 4 + 5 + 2 = 36$. This is hardly a surprise, as the sum of all the frequencies must be the total number we started with; in other words $\Sigma f = n$.

Going back to our formula for $\bar{x}$, we require the sum of terms $f_1 x_1, f_2 x_2$, etc. We can therefore say that

$$\bar{x} = \frac{\Sigma f x}{n}$$

where $\Sigma f x = f_1 x_1 + f_2 x_2 + \ldots$ and n is the total number of scores and is equal to Σf. It is often convenient to include the values of fx in the frequency table if we know we are going to have to calculate the mean, so that here we would have:

score (x)	frequency (f)	fx
1	0	0
2	1	2
3	12	36
4	12	48
5	4	20
6	5	30
7	2	14
8 and over	0	0
	$n = \Sigma f = 36$	$\Sigma f x = 150$

Mean score $\bar{x} = \frac{150}{36} \simeq 4\cdot17$.

Remember that, although the formula for $\bar{x}$ looks complicated, we are really doing nothing more than adding the scores and dividing the total by the number of scores.

How far has our analysis of the golf scores helped us? We started with a confusing list of figures and, by drawing up a frequency table and a frequency chart, managed to extract first a numerical and then a visual pattern. A quick glance at the frequency chart would enable us to assimilate quickly the overall distribution of scores, whereas we would have to stare at the original score card very much longer to perceive this. Not only have we simplified and clarified the information, we have also produced a "measure" of Jim's attainment at golf. The mode and the mean are two possible measures, the former of less use here, as there happened to be two modal values. If, on the next double round, the analysis is repeated and the mean score this time is only 3·42, then we can "reasonably" conclude that the golf has improved in the meanwhile. (We will return to the problem of "reasonable" conclusions from data in Chapter 12.)

4.3. Histograms: ages of cars

At the beginning of February 1970, a car salesman, interested in the possible market, decided to do a quick survey of the first 200 cars passing outside the window of his showroom to estimate their age-distribution. Because, in 1963, not all counties issued the letter A to follow their registration numbers, he combined the A and earlier registrations into one category. His results looked like this.

Letter showing year of registration	Frequency count	Frequency
A or none	HHt HHt HHt HHt HHt HHt HHt HHt HHt HHt ||||	54
B	HHt HHt HHt |	16
C	HHt HHt HHt HHt |||	23
D	HHt HHt HHt HHt ||	22
E	HHt HHt ||||	14
F	HHt HHt HHt HHt HHt |||	28
G	HHt HHt HHt HHt HHt |||	28
H	HHt HHt HHt	15

(Notice the useful trick when counting frequencies of drawing each fifth stroke crosswise.)

We could illustrate these results in the form of a frequency chart showing the frequency of each letter, exactly as we did for the golf scores. We have reversed the order of the letters since it is the age of the cars which is of interest (Fig. 4.3).

At first sight the pattern of the distribution looks rather odd—was there really a slump in car sales at the time of E and H? The explanation is that these two letters cover shorter periods of time than the others,

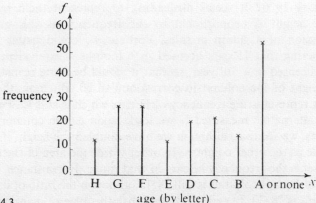

Figure 4.3

for in February 1970, H had only been issued for 6 months, and the letter E was issued only from January 1967–July 1967, as the first month of registration was then altered from January to August.

Could we adjust the chart to allow for this? We might decide to start again, this time showing age and year rather than letter as the horizontal axis of the frequency table. Of course we do not know exactly when, for instance, an F car was registered, so we will have to represent them as being spread out evenly between 1st August 1967 and 31st July 1968. We can see that a solid "bar" is more suitable than a narrow vertical line for representing the frequency. The age of cars is a **continuous** variable whereas their registration letter is discrete; as the car could have been registered at any point along the time-scale there is no real gap theoretically (although there may be an overnight one in practice), between the last registration on 31st July and the first on 1st August. Any measuring process, unlike counting, will result in a continuous scale of results.

We will first put in the diagram all the "straightforward" columns where a letter had been issued for a year as shown in Fig. 4.4.

Now, what about the cars registered between August 1969 and

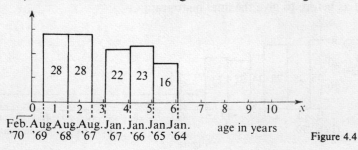

Figure 4.4

February 1970? It seems misleading to represent them by a column whose height is proportional to the frequency as this gives a false impression of a slump in sales. Perhaps we could better solve it by considering that 15 cars licensed in 6 months is equivalent to 30 cars being licensed in a full year, so that it would be more realistic to adjust the height of the column to correspond to 30. But now the height no longer represents the frequency of H cars which is still 15. Perhaps there is an alternative measure? If we look again at the column for the six months, we see that, although we have doubled the height it is only half as wide as the other columns. In other words the area of the column still represents the frequency of cars in that range. For instance, the ratio of areas of G to H columns is still 28:15, although the ratio of heights of the columns is 28:30. In order to distinguish this type of diagram, where the variable along the horizontal axis is continuous and the frequency is represented by the area of the columns, we call it a **histogram** rather than a frequency chart. Since the units of area represent frequency, and the horizontal units represent years, the units of the vertical axis can only be "frequency per year". In fact we usually label the vertical axis *frequency density*, by analogy with population density (people per hectare) or physical density (g per cm³), so that this phrase can be used whatever the horizontal units.

We can use a similar method to represent the E cars. As the width is $\frac{7}{12}$ of a year, if we let the height be h units, then the area $7h/12$ must represent the frequency 14. Hence $h = \frac{12}{7} \times 14 = 24$.

We have even worse problems with the pre-1964 cars, though, as we do not know how many years the interval covers. We have the choice of leaving them off altogether (which might be safer); doing a further survey of the individual cars in order to find a more realistic idea of their actual ages (which would be more precise but time-consuming); or of assuming a reasonable, if arbitrary, starting point for the interval. Suppose we decided on the last approach and considered it unlikely that any of the cars were registered before 1956, then we can adjust the column height as before to give the final histogram.

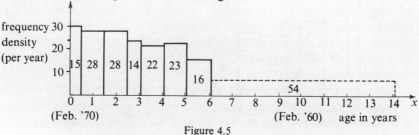

Figure 4.5

4.4. The mode and mean: ages of cars

Suppose the car salesman had carried out a similar survey of cars in a different area and wished to compare the two sets of figures. If he wanted to compare the two sets of data, it would be convenient to have them both plotted on the same piece of paper, and it is almost impossible to superimpose two histograms. In this situation a *frequency polygon* is usually plotted instead. This is simply derived from the histogram by joining the mid-points of the tops of the columns, as the figure shows. It is not as informative as the histogram (and more difficult to plot without first drawing the histogram to get the heights correct); its only advantage is that two frequency polygons can be easily compared.

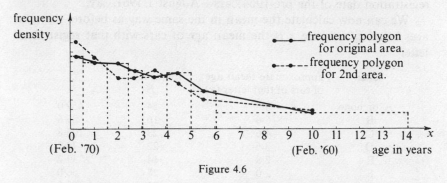

Figure 4.6

Having plotted the two distributions, the salesman might then ask us to compare the "typical" age of a car in each of the two areas. Which measure should we choose? We could take "typical" to mean "occurring most frequently" and choose the mode.

Even in the golf example we found difficulty with the mode, as there happened to be two possible values, 3 and 4. Here, however, the situation is very much worse. As we do not know any of the actual ages but only the range in which each lies, it seems that we cannot give a value for the mode but only the modal class; and as the classes have differing lengths we should take this into account and give as the modal class the one corresponding to the greatest column height of the histogram (or the highest point of the frequency polygon). Therefore for the sample we have been considering it would appear that the modal class contained those cars with ages less than 0·5 year.

In fact in any case where the data are only given in groups, the best we can ever do is to give a modal class; we can never give a single value of the mode as we may be able to do in a discrete case. It must be clear by now that although the mode, or modal class, is usually easy to pick out from the frequency chart or histogram, there are so many difficulties associated with it that it has a very limited use statistically.

Nor is it simple to calculate the mean age of the cars; we can hardly add up the ages and divide by the total number (200), as we do not know the *exact* age of any of the cars. We can however make an estimate by assuming that each car bearing a given letter was registered exactly half way through the corresponding period. Even this does not solve the pre-1964 cars; what is their average age? Strictly we ought to do a separate investigation to determine this. Let us assume however that our salesman requires only a rough estimate, so that we could without great loss take an arbitrary but reasonably likely date for the mean registration date of the pre-1964 cars—August 1 1961, say.

We can now calculate the mean in the same way as before, the total ages being Σfx, where x is the mean age of cars with that registration letter:

letter	approximate mean age of cars of that letter (x)	frequency (f)	fx
A or none	8·5	54	459·0
B	5·6	16	89·6
C	4·6	23	105·8
D	3·6	22	79·2
E	2·8	14	39·2
F	2·0	28	56·0
G	1·0	28	28·0
H	0·3	15	4·5
		$\Sigma f = 200$	$\Sigma fx \simeq 861$

Approximate mean age $\Sigma fx / \Sigma f = 861/200 \simeq 4.3$ years.

This is only a rough estimate, even for this sample. As you can see a small difference in our "guessed" mid-point for pre-1964 cars would make a considerable difference to the estimate.

In spite of all the difficulties we have encountered along the way, we have managed to extract from the original data a histogram illustrating the trend of car sales, and a rough estimate of the mean age of cars which we could use for comparing the salesman's two distributions.

4.5. Grouping data: travel of snails

Geoffrey wants to investigate the mobility of snails. He decides to take a sample of 103 snails and measure how far away from a fixed starting line they are after one hour. The measurements he obtains are given to the nearest centimetre below:

55	137	28	53	192	67	79	61	13	83
27	112	88	166	35	69	68	87	39	75
60	260	87	17	98	91	149	73	193	51
88	107	111	104	71	112	136	89	84	53
105	56	96	128	61	171	95	305	54	84
66	37	69	128	57	135	165	92	60	95
249	188	73	67	136	63	56	134	159	94
100	105	97	115	67	60	84	71	56	79
93	102	57	80	275	100	63	62	78	57
65	17	136	146	129	128	147	150	172	123
99	131	156							

In order to illustrate these results, it would hardly be feasible to draw a frequency chart with each distance having a separate frequency column, as few of the measurements occur more than once, and many intermediate ones not at all. We can make the pattern clearer by agreeing to group the lengths in certain ranges; we might for instance take every measurement between 100 and 109 cm as one group, and so on.

How do we decide which grouping to choose? If we take each interval too small we are not much better off than when we started, as we have a large number of very small columns which appear random (Fig. 4.7). On the other hand we can see the result of taking intervals that are too large (Fig. 4.8); the real shape of the distribution is obliterated.

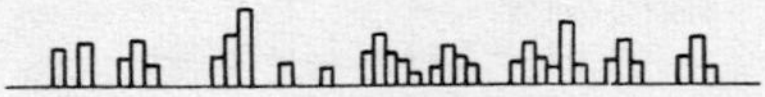

too many intervals

Figure 4.7

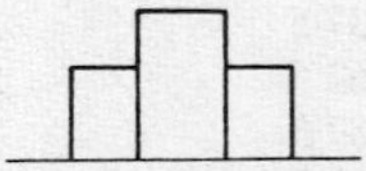

too few intervals

Figure 4.8

A reasonable compromise is to choose between ten and twenty intervals to cover all the measurements. It is useful to make all the intervals the same length and to arrange for their mid-points to be round numbers. Another good rule is to choose the intervals so that none of the values lies on the boundary of two intervals.

What about the case of the snails? We have to cater for a range of at least 288 cm, from 17 to 305 cm. This rather suggests we aim for either about 15 groupings each covering an interval of 20 cm, or about 12 with intervals of 25 cm or about 10 with intervals of 30 cm.

Suppose we decide to settle for about 12 groupings, each 25 cm long. Should we take, for instance, 10–35, 35–60, 60–85 . . . ? This gives difficulties straight away. In which class do we put the reading of 60 cm? Suppose we alter them to 10–34, 35–59, 60–84 . . . . But what about the mid-points of the intervals? They work out to be 22, 47 . . . , hardly the most convenient numbers to work with.

What mid-points would we like to work with, knowing that in this case they will increase by 25 cm each time? Clearly 25, 50, 75 . . . would be the most convenient; so suppose we attack the problem from this end and choose our group intervals round the mid-points. This gives us intervals of 13–37, 38–62, 63–87 . . . .

We notice, however, that when we are concerned with a variable that represents length or weight or time, etc., then it is continuous. In the case of the snails we are only restricted to whole numbers of centimetres in the data because measurements were taken to the nearest centimetre, and so we should take a measurement of 73 cm to mean a real distance of between 72·5 and 73·5 cm. So it is really better to re-label our intervals 12·5–37·5, 37·5–62·5, 62·5–87·5 . . . so that there are now no gaps between the intervals along the distance axis. Notice that this does not alter the mid-points of the intervals, nor does it matter that this time 62·5 comes at the end of one interval and the beginning of the next, as we have no measurements of 62·5 cm in the data.

This may seem a very complicated procedure, but it does help considerably to ease the arithmetic later on and is worth a little effort.

interval (cm)	mid-point (cm)	frequency count	frequency f
12·5– 37·5	25	﷐﷑ ‖	7
37·5– 62·5	50	﷐﷑ ﷐﷑ ﷐﷑ ‖‖	18
62·5– 87·5	75	﷐﷑ ﷐﷑ ﷐﷑ ﷐﷑ ﷐﷑	25
87·5–112·5	100	﷐﷑ ﷐﷑ ﷐﷑ ﷐﷑ ‖‖	23
112·5–137·5	125	﷐﷑ ﷐﷑ ‖‖	13
137·5–162·5	150	﷐﷑		6
162·5–187·5	175	‖‖‖	4
187·5–212·5	200	‖‖	3
212·5–237·5	225		0
237·5–262·5	250	‖	2
262·5–287·5	275			1
287·5–312·5	300			1

$$\Sigma f = 103$$

We can now draw the histogram. Although it is the areas of the columns that are proportional to frequency, as the columns are here of equal width, their heights will also be proportional to the frequency. (In fact this often happens.)

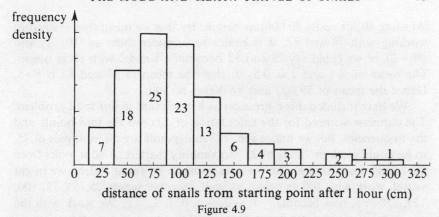

frequency density

distance of snails from starting point after 1 hour (cm)

Figure 4.9

4.6. The mode and mean: travel of snails

We have now obtained an idea of the shape of the distribution. What about the mode? As the distribution is grouped, the histogram will only tell us the modal class, which is 62·5–87·5 cm. In other words more snails travelled between 62·5 cm and 87·5 cm than between any other comparable interval. Here is yet another of the difficulties of using the mode; it is quite conceivable that if we had chosen the groupings differently we might have obtained a different modal class. Of course we could instead have gone back to the original table and found the actual measurement which was the most popular. However, we find that the values 56, 57, 60, 67, 84, 128 and 136 all occur three times, and none more often. So although the modal class is slightly suspect it does look as though it is giving us rather more information than is conveyed by the list of seven actual modal values.

What about the mean? If a calculating machine (or computer) is readily available, the most accurate and least complicated way to find the mean is simply to add up the values and divide by the total number (103 in this case).

If you are unlucky enough to have to work it out without any assistance, then it is probably rather quicker to use the Σfx formula for finding the total, where each measurement is assumed to take the value at the mid-point of its interval (which is why we chose it to be a convenient number). Naturally, because this assumption is made, the method will only give an estimate of the mean, but it should be reasonably accurate provided that the intervals were not chosen to be too long.

What is the mean of 79 000 and 82 000? The answer is obvious without adding and halving by (a) working in units of 1000 (in other words regarding the last 3 zeros as irrelevant in the working out) and

(b) using 80 (or really 80 000) as origin. By this we mean that instead of working with 79 and 82, it is easier to consider them as $(80-1)$ and $(80+2)$, or we could say 79 and 82 become -1 and 2 with 80 as origin. The mean of -1 and 2 is 0·5, so that the mean of 79 and 82 is 80·5. Hence the mean of 79 000 and 82 000 is 80 500.

We have included this digression as it is relevant to our snail problem. The columns we need for the calculation of Σfx are the mid-points and the frequencies. But we notice that the mid-points are all multiples of 25, so we could work in "units of 25" to simplify matters. Also it looks from the histogram as though the mean is going to be about 100, so we might as well work with 100 as origin as well. In other words 25, 50, 75, 100, 125, 150 . . . . now become -3, -2, -1, 0, 1, 2. . . . . We work with the new value of x and enter it in a new column.

mid-point of interval	origin 100, units of 25 x	f	fx
25	-3	7	-21
50	-2	18	-36
75	-1	25	-25
100	0	23	0
125	1	13	13
150	2	6	12
175	3	4	12
200	4	3	12
225	5	0	0
250	6	2	12
275	7	1	7
300	8	1	8

$$\Sigma f = 103 \qquad \Sigma fx = 76-82$$
$$= -6$$

The estimated mean (with origin 100 and in units of 25 cm) is $(\Sigma fx)/n = -6/103$.

Going back to units of 1 cm, the estimated mean (origin 100) $= \frac{-6}{103} \times 25 \simeq -1·5$.

∴ estimated mean distance travelled is $100-1·5 = 98·5$ cm.

You will probably now see the necessity for calculating machines and computers in statistics, as other methods can be rather tortuous as well as less accurate. In fact the loss of accuracy in this case is very small, because a large number of groupings were used. The smaller each interval is, the smaller the loss of accuracy in assuming each measurement to lie at the mid-point of its interval.

You are free, of course, to use any method you wish to calculate the mean; without a machine the method that is shown above involves least arithmetic, but if you would rather not work in terms of different units and a new origin, then this really does not matter.

With a machine, you will probably have little difficulty in drawing up a flow chart for the calculation of the mean from the original data (Fig. 4.10):

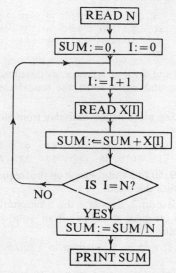

Figure 4.10

This can be easily modified to calculate the mean from a frequency table using the $(\Sigma fx)/n$ formula.

We now have a method for classifying and analysing a mass of data by, (i) grouping into suitable class intervals if necessary (of course a discrete distribution with a large number of different values can also be grouped in the same way to simplify calculations), (ii) drawing up a frequency table, (iii) drawing a histogram or frequency chart, (iv) working out the mode or modal class if it is convenient and meaningful to do so, (v) working out the mean (or an estimate of it).

EXERCISE 4

1. The table below gives the frequency of occurrence of values of x within certain ranges. Draw a histogram of the distribution.

x	frequency	x	frequency
1–2	3	5–6	55
2–3	10	6–7	26
3–4	21	7–8	15
4–5	45		

These frequencies may be regarded as relating to the occurrence of values of x^2, so that in the range 1–4 there are 3 values of x^2, in the range 4–9 there are 10 values, in the range 9–16 there are 21 values, etc. Draw a histogram of the distribution of these frequencies. Comment on the difference between the shapes of the two histograms.   (J.M.B.)

2. The marks of 50 candidates in an examination for which the maximum mark was 100 are given below:

62	21	4	26	7	38	32	64	12	38	45	6	33	55	62
48	49	7	9	41	21	30	31	3	25	57	48	48	18	49
72	23	5	8	37	31	31	39	65	53	4	75	17	14	61
50	51	38	36	40										

Select suitable classes and draw up a frequency distribution. Draw a histogram to represent the data and write down the modal class. Comment on any significant features.   (Cam.)

3. The number of sparking plugs issued each day from stock in a garage for the 30 days of a month are:

| 75 | 72 | 81 | 91 | 101 | 108 | 61 | 72 | 83 | 92 | 74 | 83 | 76 | 62 | 100 |
| 99 | 98 | 76 | 58 | 73 | 102 | 56 | 68 | 57 | 83 | 81 | 79 | 102 | 97 | 69 |

Using groups 55 to 59, 60 to 64, etc., draw up the frequency distribution table. From this table estimate the mean.   (Cam.)

4. Using the data of question 2, calculate the mean mark of the 50 candidates (i) directly from the original data, (ii) from your frequency distribution. Compare the two results.

5. The number of runs scored by a batsman in 9 successive completed innings were:

$$56 \quad 12 \quad 34 \quad 82 \quad 6 \quad 4 \quad 63 \quad 0 \quad 1$$

What would he need to score in the next innings to give him an average of 30?

6. A record was kept of motor vehicles passing in one direction along a main road. The number of private cars in each group of 10 vehicles was noted. When 60 successive groups had passed the frequencies were as follows:

No. of private cars in a group of 10	0	1	2	3	4	5	6	7	8	9	10
No. of groups	0	0	4	5	8	13	11	10	7	2	0

(a) Calculate the mean number of private cars in a group of 10 vehicles.
(b) Estimate the probability, at a given instant during the census, that the next vehicle to pass would be a private car.
(c) Taking the mean length of a private car to be 5 metres and of a commercial vehicle to be 7 metres, and allowing 9 metres between vehicles, calculate the length of road occupied by the modal group of vehicles.
(d) Represent the information in the above table in a suitable graphical form.
                                        (S.M.P.)

7. Working with an origin of 3·6 years and in units of 0·1 of a year, recalculate the mean age of the cars in the investigation on page 50 and verify that you obtain the same result.

8. An analysis of the number of words per sentence in the first 100 sentences of (a) Pride and Prejudice by Jane Austen, (b) The Cathedral by Hugh Walpole, gives the following frequency table:

No. of words per sentence	No. of sentences (a)	(b)	No. of words per sentence	No. of sentences (a)	(b)
0–4	6	2	55–59	0	1
5–9	33	12	60–64	0	2
10–14	22	14	65–69	0	0
15–19	15	19	70–74	0	1
20–24	10	18	75–79	1	0
25–29	4	6	80–84	0	0
30–34	2	5	85–89	0	0
35–39	3	6	90–94	0	1
40–44	2	6	95–99	0	0
45–49	2	4	100–104	0	2
50–54	0	1			

On one sheet of graph paper draw two frequency polygons, setting them out in such a way that you can use them to compare the two distributions gives above. Discuss the use of this analysis to demonstrate that one of the novels is more difficult to read than the other. (J.M.B.)

9. With the data of the last question find as accurately as you can the mean number of words per sentence for each of the two books and write down the modal class in each case.

Cumulative Frequency and the Median

5.1. Introduction

We drew frequency charts and histograms in the last chapter to illustrate diagrammatically the frequency distributions of three variables: the number of strokes per hole of miniature golf, the age of a car, and the distance travelled by a snail in an hour. These would enable us to answer quickly questions like: "How many times was a hole done in 3 strokes?" or "How many snails travelled between 63 and 87 cm in the hour?"

However we are not always primarily concerned with the actual frequency of each value (or range of values). For instance if we have a set of examination marks, which are all whole numbers out of 100 and which we group in fives, we might not be so interested in the question: "How many people had marks between 20 and 25?" as in the questions: "How many people had marks less than or equal to 20?" or "How many of the marks were less than or equal to 45?" This would be particularly so if 46 were the pass mark, or alternatively if we could not decide whether to set the pass mark at 46 or 51 and wanted to see how many would fail in each case. In such cases it is often useful to draw up a second table recording the *total* number scoring less than or equal to 5, 10, 15, . . . , and so on, so that we can read at a glance the numbers gaining no more than any given multiple of 5 marks.

Because we are in each case adding on successive frequencies, e.g. the total number less than or equal to 20 is obtained by adding the frequency of the 16–20 group to the number less than or equal to 15, the resulting table is called a **cumulative frequency distribution** or table. A diagram can then be drawn to illustrate the cumulative frequencies, as will be seen in the particular cases considered in the next two sections, each of which uses data given in the previous chapter.

5.2. Cumulative frequency: golf scores

The original data for the double round of golf are shown in the table.

x	f
1	0
2	1
3	12
4	12
5	4
6	5
7	2
8 and over	0

This quickly answers the question: "On how many holes were exactly 5 strokes taken?" However, the questions that are most likely to interest us are: "How many scores were good ones?", "How many could be called satisfactory?" and "How many were bad ones?" Of course in each case the standards of "good" and "bad" are entirely at our discretion; let us suppose that we count a hole which was done in 3 strokes or less as a good one, one in which 4 or 5 strokes were taken as satisfactory, and a hole with more than 5 as bad.

We can draw up a full cumulative frequency table showing how many scores were less than or equal to a particular number.

x	F	
1	0	
2	1	
3	$1+12 = 13$	
4	$1+12+12 = 13+12 = 25$	F = cumulative frequency
5	$1+12+12+4 = 25+4 = 29$	(total number of scores $\leqslant x$)
6	$1+12+12+4+5 = 29+5 = 34$	
7	$1+12+12+4+5+2 = 34+2 = 36$	
8	$1+12+12+4+5+2 = 36$	

There are one or two points to notice from the table above: first, that the easiest method of obtaining the cumulative frequency, say F_5 for the value 5, is by adding f_5 to F_4, in other words simply adding on the corresponding value from the original frequency table at each stage. However, it is still true that $F_5 = f_1 + f_2 + f_3 + f_4 + f_5$. Instead of writing this out we could use the Σ notation again, meaning that we add up the values of f. However, as in this case we do not want to add up *all* the values of f, only the ones for values of x from 1 to 5, we can write:

$$F_5 = \sum_{i=1}^{5} f_i,$$

where we interpret the right-hand side to mean "the total of all the frequencies f_i where i takes all the values from 1 to 5 inclusive". (We shall continue to leave out the subscripts if the summation is over all values of i.)

Of course at this stage a formula is hardly necessary, but it is useful to mention it as we can compare it with other formulae in later chapters.

The second point to notice is that the cumulative frequency always increases (or stays the same) as x increases, for each time we are adding a non-negative number to the previous value. As we come to the maximum value of x we are bound to have the corresponding value of F equal to the total number in the sample; for instance in this case $F_7 = 36$ as all scores are $\leqslant 7$. We also notice that we have $F = 0$ for any score less than the minimum.

We can now use the cumulative frequency table to say that the number of good holes (3 strokes or less) is 13; the number of satisfactory holes (4 or 5 strokes) is $29 - 13 = 16$, and the number of bad holes (more than 5 strokes) is $36 - 29 = 7$.

We can draw a **cumulative frequency chart** to illustrate this distribution in much the same way as the frequency chart illustrated the frequency distribution. Though in this case it might not seem necessary to join up the separate points, we can do so if we agree that, for instance, the number of scores $\leqslant 4.5$ is equal to the number of scores $\leqslant 4$ and so on. The graph moves upwards in a series of jumps at each of the possible scores and is known as a *step function*.

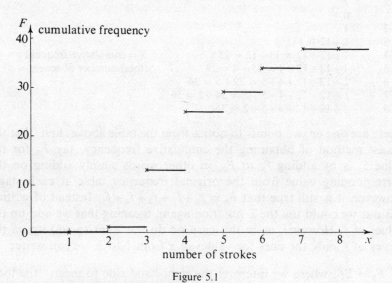

Figure 5.1

5.3. The median: golf scores

In Chapter 4 we met two measures of the "typical" value of a variable; we could obtain the mode (the value occurring with the greatest frequency) or the mean (the average of all the values taken by the variable).

Suppose we now rephrased the question about the typical golf score to read, "What was the score for the average hole?" At first glance we might agree that this was the same as the mean, but it would hardly make sense to say that the score for the average hole was 4·17, as this score is impossible for any hole. So can we distinguish between "the average score for a hole"(the mean), and "the score for the average hole"? For another example let us go back to examination marks and ask the equivalent question, "Is the average mark per child the same as the mark of the average child?" But who is the average child? Not the child who scores the mean mark, as this is probably not a whole number and therefore it is unlikely that there is any child who scores the mean mark. Is there any other way of pointing out the average child in the class as indicated by the examination marks? Most people would say that the average child was the one in the middle of the class; if there were 31 children it would be the one in the 16th position according to examination mark.

"The mark of the average child" we could therefore interpret as meaning the mark of the child who came 16th out of 31. This gives a number which is probably different from the mean; we shall call it the **median** mark M. We need not, of course, bring the children into it at all but simply arrange the marks in order and select the middle one.

You have probably foreseen one difficulty. What would have happened if we had had 30 children instead of 31? Of course there is now no "middle child" and hence no "middle mark". However we can get over this by saying that the imaginary "middle child" would come between the 15th and the 16th, and if we take the mean of the 15th and 16th marks we should arrive at a reasonable compromise. (Now the median mark is no longer necessarily a whole number.) However if the number of children was fairly large—say 100 or more—then it would be sufficiently accurate to take the median mark as the mark of the 50th child (rather than the mean of the 50th and 51st).

Suppose we apply this to the golf scores. What is Jim's score for the average hole? In other words if we arrange all the 36 scores in order of size, what is the mean of the 18th and 19th? Looking at the cumulative frequency table we see that 13 scores are less than or equal to 3, and 25 less than or equal to 4, so that both the 18th and 19th must be exactly 4, and hence the median is also 4.

We see that the median is fairly close to the mean in this case. This will always happen when the frequency chart is reasonably symmetric, and in such cases therefore the median provides a good measure of the "centre" of the distribution, while being much easier to calculate than the mean.

5.4. Cumulative frequency and the median: ages of cars

The same sort of situation arises with the cars; the salesman is more likely to want to know how many of the cars are more than 10 years old (an almost impossible figure to estimate from the frequency table) or less than 4 years old, and so on, than to know how many of the cars were between 2·5 and 3·1 years old. This will be especially true if he is investigating the market for new cars or wondering about his stock of spares. In order to answer these questions we can draw up a cumulative frequency table for the ages of cars in a similar way to the golf scores.

However this time we have to be a little more careful with the end-points of the intervals. For instance, as we are counting up the total number of cars less than or equal to a certain age, then, referring to the histogram on page 48, we have 15 cars $\leqslant 0·5$ year, $15 + 28 = 43 \leqslant 1·5$ years and so on. It is usually better to keep the frequency and cumulative frequency tables completely separate to avoid confusion between end-points and mid-points of intervals.

Age x	F	
0	0	
0·5	15	
1·5	$15 + 28 = 43$	
2·5	$43 + 28 = 71$	
3·1	$71 + 14 = 85$	
4·1	$85 + 22 = 107$	F = total number of cars whose age $\leqslant x$.
5·1	$107 + 23 = 130$	(You should contrast these x-values with
6·1	$130 + 16 = 146$	those used on page 50 for the calculation
14(?)	$146 + 54 = 200$	of the mean.)

In Fig. 5.2 we have joined together the points to form a curve, as the age of cars is a continuous variable and we would expect the F-function to increase continuously between the points which we actually can plot on the graph. The latter half of the curve is dotted, as it is very much an estimate of where the curve might lie, there being no points of reference for ages greater than 6·1 years. Where a cumulative frequency graph is continuous, it is sometimes called an **ogive**.

We can now answer the salesman's questions directly from the graph. For instance, an estimate of the number of cars over 10 years old might be about $200 - 180 = 20$, but it should be appreciated that this is an extremely approximate result. We can be more accurate though about the number of cars less than 4 years old and read from the graph that the number will be about 105. We could also have obtained this value from the cumulative frequency table by linear interpolation, which means

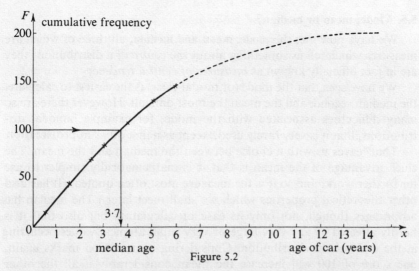

Figure 5.2

inserting a new point proportionately between two known ones. For example, as there are 85 cars less than 3·1 years old and 107 less than 4·1 years old, there will be roughly $85 + \frac{9}{10}(107 - 85) = 105$ less than 4 years old.

We can also see the pattern of the distribution from the curve. It does rise steeply at first and then rather more slowly, so that we can tell that the rate of car registration has been greater in recent years.

What about the median age? We have two possible methods of finding it. The quicker, now we have drawn the ogive, is to say that the middle car is roughly the 100th (to be exact 100·5 but the difference is clearly negligible in this case); and from the curve, if we take 100 on the vertical axis and then read down to the horizontal axis, we discover that there are 100 cars whose age is less than 3·7 years. In other words the 100th car is approximately 3·7 years old. In this case there is a considerable difference between the mean (4·3 years) and the median, as the distribution was certainly not symmetric.

We could, without drawing the curve, also estimate the median by linear interpolation, this time working in reverse. As there are 85 cars less than 3·1 years old, and 107 less than 4·1, there must be roughly 100 less than $3·1 + (100 - 85)/(107 - 85) \times 1 = 3·8$ years old. The two ways of estimating the median will, as here, give nearly the same result. Any differences will be due either to the fact that using linear interpolation is equivalent to joining the two known points with a straight line as an approximation to that part of the cumulative frequency curve, or to the difficulty of reading a value off the graph with any great accuracy.

5.5. Mode, mean or median?

We have now met the **mode, mean and median,** all three of which are measures which tell us something about the *centre* of a distribution; they are in fact officially known as *measures of central tendency*.

We have seen that the mode (or modal class) is the easiest to calculate, the median second, and the mean the most difficult. However there are so many difficulties associated with the mode, for example bimodal distributions, that it is very rarely used except as a first quick approximation.

That leaves us with a choice between the median and the mean. The chief advantage of the mean is that it is mathematically simpler to use for further work, and so it is the measure most often quoted. (It has also other theoretical properties which we shall meet later.) The median has advantages though, not only its ease of calculation but also that it is hardly affected by any very large or very small atypical values occurring at the ends of the distribution. Considering examination marks again, one score of 100 will increase the mean considerably if all the other marks are in the range 40–60, while it will make no difference to the median whether the highest mark is 60 or 100. In this case the median gives a truer picture of the centre of the distribution than the mean.

Which of the two we choose will thus depend on the sort of data with which we are dealing and the use to which we want to put the measure.

EXERCISE 5

1. The 74 golfers who qualified to play in the British Open Tournament in 1970 had the following scores for the two qualifying rounds.

score	133	136	137	138	139	140	141	142	143	144	145	146	147	148	149
frequency	1	1	2	1	3	2	5	9	18	10	5	3	2	5	7

   Find the median score required to qualify. Draw a cumulative frequency graph of the data.

2. The following table shows the number of candidates c scoring m marks for a question in an examination.

(m)	0	1	2	3	4	5	6	7	8	9	10
(c)	6	24	88	108	110	112	90	69	38	21	4

   Calculate the mean and median of the distribution. What feature of the distribution causes the mean to be greater than the median?   (Cam.)

3. The 40 completed innings in a cricket match gave the following distribution of scores.

score	0–9	10–19	20–29	30–39	40–49	50–59	60–69	70–79	80–89	90–99
frequency	10	6	7	7	3	2	2	2	0	1

   (a) Draw the cumulative frequency curve.
   (b) Estimate the median
      (i) from the curve, (ii) by linear interpolation.

(c) Estimate the number of scores which were:
(i) less than 35, (ii) greater than 55, (iii) between 13 and 73 inclusive.

4. The following numbers are the marks obtained by 50 boys in an examination.

```
57  60  37  74  62  40  56  59  80  60  62  94  78  73  56  68  67
79  83  87  90  93  58  46  77  63  66  66  56  71  51  77  53  69
70  69  70  70  47  54  49  54  68  35  64  67  76  73  68  61
```

(i) Reduce these marks to a frequency distribution, with equal intervals, having as the first interval 35–44 inclusive. Draw a cumulative frequency graph for this distribution.
(ii) What is the median mark?
(iii) Use your graph to estimate what % of candidates pass the examination, if the pass mark is 55.   (St. Dunstan's)

5. Use the table on page 52 to draw up the cumulative frequency distribution for the distances travelled by the snails, and plot the corresponding ogive. Find:
(a) the median distance travelled,
(b) the number of snails which travelled more than 100 cm,
both from your graph and directly from the original data on page 51.

Measures of Spread

6.1. The range

Suppose a man announced that he had three unmarried sisters whose mean age was 20. It might come as rather a shock to his male friends, who had at last succeeded in gaining an introduction to three such eligible young ladies, to meet two girls of 10 and 12 and a woman of 38.

So although the mean is probably our most useful measure of the centre of a distribution (in this case, the median 12 would hardly be a better measure, and there is no meaningful value for the mode) it leaves a lot to be desired in giving all the necessary information about a distribution. It certainly tells us where the average lies, but gives us no idea of how close to the average the values are. We must be told a measure of the spread of the numbers in order to differentiate between the following sets of ages for the sisters, all of which satisfy the condition of having mean 20: 10, 12, 38; 15, 22, 23; 20, 20, 20.

Probably the most obvious measure to take is the actual spread of the ages. The first trio have mean 20 but spread 10–38, whereas the third trio have mean 20 but spread 20–20. This does not give us a single measurement, but we can prune it down even more by calculating the difference between the largest and smallest measurements, which we shall call the **range.** We can now characterize the three sets of ages given above, by saying that they all have mean 20 but the ranges are 28 ($=38-10$), 8 and 0 respectively.

The range is clearly a very simple and quick-to-calculate measure of the spread of the values. However, if we return to our example about the ages of cars, we would be rather stuck for a value for the range, as, although we know how many of the cars were registered before December 1963, we do not know exactly what age the oldest one was. If we know that one of the cars was a vintage car manufactured in 1920, then we can work out the range to be $50-0 = 50$ years. We can now describe the

66

ages of cars in the sample to have mean 4·3 years with range 50 years. But is the range really a good measure of the "dispersion" (or spread) of the ages of the cars? Out of the 200 cars, probably 199 were registered between 1950 and 1970, giving a range of only 20 years. It is the single exceptional car which increases the range by 30 years, although this car is most unrepresentative of the population we are considering. Thus, although the range is very quick to calculate, it has the grave disadvantage of being determined by only two values, no matter how large the size of the sample, and that these two values are quite likely to be "freaks" in the sense that they may be much larger or smaller than the great majority of the values. Again, as with the mode, it appears that the measure most easy to obtain is in fact of little use, and that in order to arrive at something more worth while a little more labour is needed.

6.2. The interquartile range

As the range is dependent on the two extreme measurements, perhaps it could be adjusted to give a more reliable measure of the dispersion of the majority of values if the "freak" measurements were discarded. But how many so-called "freaks" are there? Could we just discard the largest five and the smallest five and find the range of the rest? Or perhaps we could leave out the top 10% and bottom 10%? There is clearly no one "right" answer to this, although a given percentage is more relevant than a fixed number, as it takes the sample size into consideration. In practice the usual solution is to leave off the top and bottom quarters of the distribution and to take the range of the remaining values. The lower point, chosen so that one quarter of the values are less than or equal to it, is called the **first quartile** Q_1. The upper point, with one quarter of the values greater than it, is the third quartile Q_3. The difference between the two is thus known as the **interquartile range**. (The median M could alternatively be called the second quartile Q_2.)

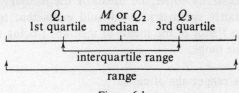

$$Q_1 \qquad M \text{ or } Q_2 \qquad Q_3$$
1st quartile   median   3rd quartile

interquartile range

range

Figure 6.1

The interquartile range is therefore the size of the interval in which the central half of the values lie. Alternatively the **semi-interquartile range** which, as we would expect, is half the interquartile range may be quoted.

6.3. Interquartile range: golf scores

Suppose we return to our example of the miniature golf scores. The cumulative frequency table is:

score (x)	cumulative frequency (F)
1	0
2	1
3	13
4	25
5	29
6	34
7	36
8	36
	36

The range is $7-2 = 5$, but again we see that the score of 2 is an exceptional one, as is perhaps the 7 also, and therefore the range does not reflect clearly the majority of the scores.

If we divide the scores in order to form sections with nine scores in each, then we can see that the first quartile Q_1 comes between the 9th and 10th scores, the median Q_2 (as we saw before) between the 18th and 19th, and the third quartile Q_3 between the 27th and 28th.

Both the 9th and 10th scores are 3, giving the first quartile as 3, and conveniently both the 27th and 28th are 5, so that 5 is the third quartile. (If, for instance, the 9th and 10th scores had been 3 and 4 respectively, we could have taken the first quartile as $3\frac{1}{2}$, in spite of the fact that it is an impossible score.)

The golf scores therefore have a median of 4 with an interquartile range of $5-3 = 2$. This tells us that the scores are centred at 4 with half of the shots lying within a range of 2 shots. Notice that the centre of this range is not necessarily either the mean or the median, although if the distribution is fairly symmetric we would guess that roughly half the scores lay within 1 shot of the mean or median, 1 being the value of the semi-interquartile range.

6.4. Interquartile range: age of cars

As we have already seen, the range of the car ages is almost impossible to work out without knowing the exact ages of the oldest and newest cars, and even then it gives little information about the age variation of the majority of the cars. However we can evaluate the interquartile range, either by linear interpolation from the cumulative frequency table (as we did for the median) or, more easily, from the ogive which we drew in the previous chapter.

The quartiles are approximately the ages of the 50th and 150th cars. Reading from the graph these are roughly 1·7 and 6·5 years, giving an

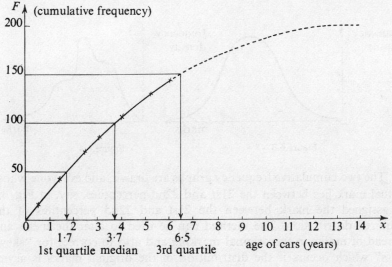

Figure 6.2

interquartile range of 4·8 years. (Note that the 6·5 years is only very approximate.) So half the 200 cars were bought within a period of about 4·8 years; and their ages lie within about 2·4 years of the median of 3·7 years. (As the distribution is not very symmetric, this last result is not very accurate, the median being rather nearer the first quartile than the third.)

6.5. Percentiles

In addition to splitting the values into four quarters whose boundary values are the quartiles and the median, the distribution is also often split into 10 parts separated by the *deciles*, the first decile being the value below which one tenth of the measurements lie, the second being that below which two tenths lie, and so on.

A very large sample may even be split into *percentiles*, where, for instance, the 23rd percentile is that value below which 23% of the measurements lie. Of course the quartiles Q_1 and Q_3 are the 25th and 75th percentiles respectively, and the median is the 50th percentile. The percentiles are usually read off from a cumulative frequency graph in the same way as the median and quartiles, and are especially useful when marks are being standardized. For example, suppose it was required that the marks for a public examination had a certain distribution, say the one in Fig. 6.3, whereas the actual marks obtained produced the distribution in Fig. 6.4.

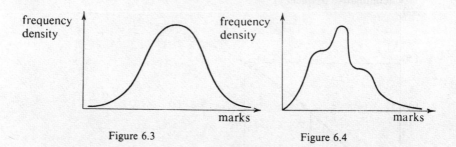

Figure 6.3   Figure 6.4

The two cumulative frequency graphs are drawn, and everyone whose actual mark lies between the 21st and 22nd percentiles, say, in Fig. 6.4 is assigned the mark between the 21st and 22nd percentiles of the required distribution. The method can be used to alter the mean and spread of marks in the original marking and also to correct for "skewness" which occurs if the distribution of the original marks is asymmetric.

EXERCISE 6A

1. In a page of a book the number of words on each line was counted, with the following results.

No. of words in a line	9	10	11	12	13	14	15	16
No. of lines	1	5	7	12	6	7	0	2

Represent these facts in a cumulative frequency graph. Find the median and quartiles of the distribution.

2. Using the data of Exercise 5, question 1, find the semi-interquartile range of the golf scores. Find also the 40th and 60th percentiles.

3. The table shows the number of men, out of a sample of 1000, with heights in the given ranges. (168— means 168 and less than 170, 170— means 170 and less than 172, and so on.)

Height (cm)	Number of men	Height (cm)	Number of men
168 —	12	182 —	136
170 —	35	184 —	98
172 —	82	186 —	66
174 —	105	188 —	40
176 —	148	190 —	12
178 —	141	192 —	nil
180 —	125		

Draw a cumulative frequency (ogive) curve to illustrate the given data. From your curve estimate:

(a) the median height,

(b) the upper and lower quartile heights,

(c) the number of men in the sample with height 183 cm or more. (Cam.)

4. The table gives details of the salaries earned by 991 employees of an insurance company.

Annual salary (£)	No. of salaries	Annual salary (£)	No. of salaries
500 –	16	1100 –	214
600 –	24	1200 –	205
700 –	55	1300 –	116
800 –	49	1400 –	42
900 –	94	1500 –	18
1000 –	158	1600 –	nil

Draw a cumulative frequency curve to illustrate the given data. From your curve estimate:

(a) the median salary,

(b) the upper and lower quartile salaries,

(c) the number of employees earning salaries between £750 and £1250 per annum. (Cam.)

5. Find the median and semi-interquartile range of the set of observations x^2 in the cases:

(a) x takes positive values only, $Q_1 = 3, Q_2 = 7$ and $Q_3 = 10$,

(b) the distribution of x is symmetric with 50th percentile zero, $62\frac{1}{2}$th percentile 3, 75th percentile 5, and $87\frac{1}{2}$th percentile 7.

6.6. The standard deviation

If we return to our definition of the semi-interquartile range as half the distance between the quartiles, we see that in a reasonably symmetric distribution it corresponds roughly to the difference between a quartile and the median. We can take this as being the "average" difference between a measurement and the median, as about half the differences should be more than this and half of them less. The **semi-interquartile range** is thus a rough measure of the median deviation of values from the median.

Can we not similarly find the average deviation from the mean and use this too as a measure of dispersion?

Suppose we return to our example at the beginning of the chapter of the three sisters with mean age 20; if their actual ages were 10, 12 and 38, then we can draw up a table as follows:

age (x)	mean $(\bar{x})$	deviation $(x - \bar{x})$
10	20	−10
12	20	− 8
38	20	18

The average deviation is thus $(-10-8+18)/3 = 0$. Is this what we want? If we think about it, we have in fact chosen the mean so that the deviations of the other measurements from it balance out to 0, taking the sign into consideration.

Surely what we want here is the numerical size of each deviation without taking the sign into account. Let us try again:

age	mean	deviation
(x)	$(\bar{x})$	$|x-\bar{x}|$
10	20	10
12	20	8
38	20	18

($|x-\bar{x}|$ is called the **modulus** or mod of $(x-\bar{x})$ and is equal to the numerical value of $(x-\bar{x})$ without considering the sign, e.g. $|-7| = |+7| = 7$.) The mean deviation is now $\dfrac{\Sigma|x-\bar{x}|}{n} = \dfrac{10+8+18}{3} = 12$, which gives a reasonable answer. So we could characterize the distribution of ages by saying that it has mean 20 and **mean deviation** 12.

The *mean deviation* is a perfectly respectable measure of dispersion (in fact it is sometimes taken about the median rather than the mean), but there are some disadvantages when it is considered from a mathematical point of view. The modulus function is a notoriously difficult one to handle mathematically. So we look to see if there is any other way in which we can make all the deviations positive without simply ignoring their sign. Probably the first way which suggests itself is the operation of squaring; whatever the sign of the deviation, we obtain a positive sign if we square it. We can therefore obtain the *mean squared deviation,* which is called the **variance** as follows:

age	mean	deviation	squared deviation
(x)	$(\bar{x})$	$(x-\bar{x})$	$(x-\bar{x})^2$
10	20	-10	100
12	20	-8	64
38	20	18	324
			488

$$\text{Variance} = \frac{\Sigma(x-\bar{x})^2}{n} = \frac{488}{3} \simeq 162 \cdot 7$$

If we look at the units of the variance we see that, as the original measurements were in years, the variance is in "square years", which is a rather odd unit for a measure of spread. We can however easily correct this by taking the square root of the variance. This measure is known as the **standard deviation** which we shall denote by S. It is measured in the same units as the mean and all the other values.

So we can say of the standard deviation:

$$S = \sqrt{\text{variance}}$$

$$= \sqrt{\frac{\Sigma(x - \bar{x})^2}{n}}$$

where $\bar{x}$ is the mean and n is the number of measurements.

In the case of the sisters' ages, $S \simeq \sqrt{162 \cdot 7} \simeq 12 \cdot 8$; so here the standard deviation gives much the same result as the mean deviation.

It seems to be one of the injustices of life that the most frequently used measure of dispersion (the standard deviation) is, like the mean (which is the most-used measure of central tendency) by far the most difficult to calculate. However it does have the advantage of fitting into theoretical probability theory very nicely, and it can be redeemed a little by using simplified methods of calculation.

6.7. Easier methods of calculating the standard deviation

To show a practical evaluation of the standard deviation, we will return again to the scores at miniature golf. We have already calculated the mean $\bar{x}$ to be $4 \cdot 17$ (see p. 45).

We could start as in the last example by calculating all the deviations, noting that each value of the deviation is repeated several times according to the frequency with which the corresponding score occurs.

score (x)	frequency (f)	deviation $(x - \bar{x})$	squared deviation $(x - \bar{x})^2$	$f(x - \bar{x})^2$
1	0	$-3 \cdot 17$	$10 \cdot 05$	0
2	1	$-2 \cdot 17$	$4 \cdot 71$	$4 \cdot 71$
3	12	$-1 \cdot 17$	$1 \cdot 37$	$16 \cdot 44$
4	12	$-0 \cdot 17$	$0 \cdot 03$	$0 \cdot 36$
5	4	$0 \cdot 83$	$0 \cdot 69$	$2 \cdot 76$
6	5	$1 \cdot 83$	$3 \cdot 35$	$16 \cdot 75$
7	2	$2 \cdot 83$	$8 \cdot 01$	$16 \cdot 02$
$\geqslant 8$	0			

$$\Sigma f(x - \bar{x})^2 = 57 \cdot 04$$

$$S = \sqrt{\frac{\Sigma f(x - \bar{x})^2}{n}} \simeq \sqrt{\frac{57 \cdot 04}{36}} \simeq \sqrt{1 \cdot 58} \simeq 1 \cdot 26$$

We see that each squared deviation is multiplied by the frequency with which it occurs. This makes the full formula $S = \sqrt{\dfrac{\Sigma f(x - \bar{x})^2}{n}}$ for frequency distributions. (The previous section gave a simplified form in which each value of x occurred only once and so $f = 1$ throughout.)

We can see that the calculation of the squared deviations is most involved if the mean is not a whole number, when each value to be squared will have some digits after the decimal point. Suppose we look again at the formula to see whether it can be simplified.

$$\Sigma f(x-\bar{x})^2 = f_1(x_1-\bar{x})^2 + f_2(x_2-\bar{x})^2 + f_3(x_3-\bar{x})^2 + \ldots$$
$$= f_1(x_1^2 - 2x_1\bar{x} + \bar{x}^2) + f_2(x_2^2 - 2x_2\bar{x} + \bar{x}^2)$$
$$+ f_3(x_3^2 - 2x_3\bar{x} + \bar{x}^2) + \ldots$$

If we gather the terms together we have:

$$\Sigma f(x-\bar{x})^2 = (f_1x_1^2 + f_2x_2^2 + f_3x_3^2 + \ldots) + (-2f_1x_1\bar{x} - 2f_2x_2\bar{x}$$
$$- 2f_3x_3\bar{x} - \ldots) + (f_1\bar{x}^2 + f_2\bar{x}^2 + f_3\bar{x}^2 + \ldots)$$

But $-2\bar{x}$ is a common factor in the second bracket and $\bar{x}^2$ in the third so
$$\Sigma f(x-\bar{x})^2 = \Sigma fx^2 - 2\bar{x}\Sigma fx + \bar{x}^2\Sigma f.$$

So $S^2 = \dfrac{\Sigma f(x-\bar{x})^2}{n} = \dfrac{\Sigma fx^2}{n} - \dfrac{2\bar{x}\Sigma fx}{n} + \dfrac{\bar{x}^2\Sigma f}{n}$

But $\dfrac{\Sigma fx}{n} = \bar{x}$ and Σf = total number of measurements = n.

Hence $S^2 = \dfrac{\Sigma fx^2}{n} - 2\bar{x}.\bar{x} + \bar{x}^2 = \dfrac{\Sigma fx^2}{n} - \bar{x}^2.$

This gives us an alternative formula for S^2 (and hence for S), which does look rather easier to evaluate as it only involves working out Σfx^2 rather than $\Sigma f(x-\bar{x})^2$.

$$S^2 = \frac{\Sigma fx^2}{n} - \bar{x}^2$$

If we set out the golf scores again for the calculation both of $\bar{x}$ and S we have:

score (x)	frequency (f)	fx	fx^2
1	0	0	0
2	1	2	4
3	12	36	108
4	12	48	192
5	4	20	100
6	5	30	180
7	2	14	98
8	0	0	0
	$\Sigma f = 36$	$\Sigma fx = 150$	$\Sigma fx^2 = 682$

$\bar{x} = \dfrac{\Sigma fx}{n} = \dfrac{150}{36} \simeq 4{\cdot}17$ (as we obtained in Chapter 4, page 45)

$S^2 = \dfrac{\Sigma fx^2}{n} - \bar{x}^2 = \dfrac{682}{36} - \left(\dfrac{150}{36}\right)^2 \simeq 18{\cdot}94 - 17{\cdot}36 = 1{\cdot}58$

$S \simeq 1{\cdot}26$

(note that this gives the same answer as before).

We can simplify our working even more by "working with a new origin". By looking at the table we guess that the mean will be roughly 4, and so we measure all the values with 4 as origin.

score	score (x) with origin 4	f	fx	fx^2
1	-3	0	0	0
2	-2	1	-2	4
3	-1	12	-12	12
4	0	12	0	0
5	1	4	4	4
6	2	5	10	20
7	3	2	6	18
8	4	0	0	0

$$\Sigma f = 36 \quad \Sigma fx = 20 - 14 = 6 \quad \Sigma fx^2 = 58$$

$$\bar{x} = \tfrac{6}{36} \simeq 0{\cdot}17 \qquad S^2 = \tfrac{58}{36} - (\tfrac{6}{36})^2 \simeq 1{\cdot}61 - 0{\cdot}03 = 1{\cdot}58, \ S \simeq 1{\cdot}26.$$

So the mean score is 4·17 as we are measuring x from 4 as origin. As S involves deviations from the mean only, it will have the same value whatever point is taken as origin. (Try to show this by proving that if you subtract any constant from all your x-values S will be unaltered.) So when we use a new origin we have to remember (a) to adjust the final mean, (b) to substitute the mean measured *from the new origin* in the formula for S, which then gives the correct value for S.

Actually in this example you will realize that the median and semi-interquartile range are 4 and 1, whereas the mean and standard deviation are 4·17 and 1·26 respectively. Therefore, in this rather trivial exercise, the much more easily calculated median and semi-interquartile range would have been very satisfactory measures of the central tendency and dispersion, and it was hardly necessary to bother with the more careful calculations to obtain the mean and standard deviation. However, if we had wished to use the data to make conjectures about Jim's golfing prowess or to compare it with another set of similar data, the mean and standard deviation have more manageable properties.

6.8. A further simplification: travel of snails

Just to repeat the above procedure and add one further point, we will take the snail distances as our data. We will assume that each value lies at the mid-point of its interval and take 100 cm as our origin and work in units of 25 cm as we did in §4·6 (page 54).

interval	midpoint	midpoint with origin 100 and units of 25, (x)	f	fx	fx^2
12·5– 37·5	25	-3	7	-21	63
37·5– 62·5	50	-2	18	-36	72
62·5– 87·5	75	-1	25	-25	25
87·5–112·5	100	0	23	0	0
112·5–137·5	125	1	13	13	13
137·5–162·5	150	2	6	12	24
162·5–187·5	175	3	4	12	36
187·5–212·5	200	4	3	12	48
212·5–237·5	225	5	0	0	0
237·5–262·5	250	6	2	12	72
262·5–287·5	275	7	1	7	49
287·5–312·5	300	8	1	8	64

$$\Sigma f = 103 \qquad \Sigma fx = 76 - 82 \qquad \Sigma fx^2 = 466$$
$$= -6$$

As before $\bar{x} = \dfrac{-6}{103}$ but this is measured in units of 25 cm with origin 100 cm, so that the mean distance travelled is

$$100 - \frac{6}{103} \times 25 \text{ cm} \simeq 98 \cdot 5 \text{ cm}.$$

$$S^2 = \frac{\Sigma fx^2}{n} - \bar{x}^2 = \frac{466}{103} - \left(\frac{-6}{103}\right)^2 \simeq 4 \cdot 52$$

$$\therefore S \simeq 2 \cdot 1$$

(Remember to substitute $\bar{x}$ from the same origin and in the same units as x.)

We saw in the last section that we do not need to adjust S for the change of origin. But what about the change of units? Remember that S is measured in the same units as x, and is therefore here measured in our new units of 25 cm. Thus the standard deviation in centimetres travelled is $2 \cdot 1 \times 25 = 52 \cdot 5$.

It is just as well to check at the end that the answer is a reasonable size to make sure we have adjusted the units correctly.

6.9. Use of calculating machines and computers

The sort of tedious calculations involved in working out the mean and standard deviation are clear candidates for the use of machines.

On a calculating machine it is possible to perform both calculations at the same time. The flow chart for calculating a mean is shown in Chapter 4 (page 55). This can be modified to include the calculation of a standard deviation as shown in Fig. 6.5.

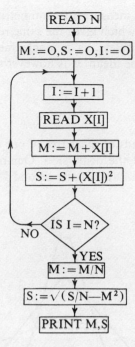

Figure 6.5

The data are read in here as single values, but the program can be modified without much effort in order to read in values from a frequency table. (In practice, if the standard deviation of a large set of numbers is to be calculated on a computer, the formula $S = \sqrt{\dfrac{\Sigma(x-\bar{x})^2}{n}}$ is often preferred as this avoids large rounding errors.)

6.10. The coefficient of variation

If we had measured the distances travelled by the snails in millimetres, we would have obtained 525 as the value for the standard deviation instead of 52·5. At first sight it would appear that the dispersion of the results is much greater with a standard deviation of 525 than with 52·5, although we know that they are exactly equivalent in this case. In order not to give rise to such mistaken impressions, the standard deviation is sometimes quoted as a proportion of the mean, which keeps it in proper perspective. This measure is called the **coefficient of variation** V, so that

$V = S/\bar{x}$ (or often $\dfrac{100S}{\bar{x}}$ when V is expressed as a percentage).

For the snails, if we measure in centimetres, $V = 52 \cdot 5/98 \cdot 5$ and in millimetres $V = 525/985$ which gives the same result.

V, which is dimensionless, is a single measure summarizing the standard deviation and the mean. It is however more informative to give both S and $\bar{x}$ individually.

6.11. Other measures

We now have two kinds of measures of our distributions; a measure of the centre and one of the dispersion. Our preferred measure of the first kind will be the mean and of the second kind will be the standard deviation.

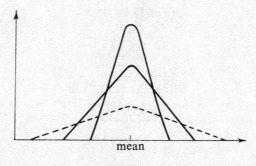

Figure 6.6

Figure 6.6 shows histograms of distributions having the same mean and varying standard deviations, and Fig. 6.7 shows some with the same standard deviation and different means. (The stepped shapes of the histograms have been rounded off in these two diagrams and in some of the following ones for the sake of clarity.)

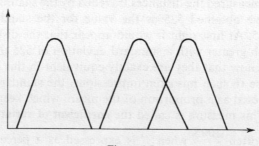

Figure 6.7

Are there any other important measures which we could use to characterize distributions? For instance, if two histograms had the same mean and standard deviation, in what ways could they differ? Figures 6.8–6.13 show varying distributions with this property.

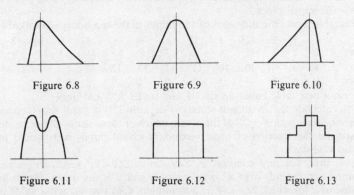

Figure 6.8   Figure 6.9   Figure 6.10

Figure 6.11   Figure 6.12   Figure 6.13

Although we could differentiate between the first three by a measure of skewness (how much the distribution is overweighted to one side or the other), it is clear that the only way to differentiate properly between the others is to know the exact shape of the distribution.

So the mean (or median) and standard deviation (or semi-inter-quartile range, or mean deviation) give a reasonable idea of the location and spread of a distribution, and this is about as much as we can use-fully find out without going into the exact details of the frequency distribution.

EXERCISE 6B

1. Travelling to work, I have two alternative routes, both of which involve queueing at traffic lights. I record, over a series of 8 journeys by each route, the time spent in queues; the results are given in the table (in minutes):

   | Route A | 15 | 15 | 16 | 10 | 17 | 20 | 14 | 13 |
   | Route B | 13 | 16 | 15 | 15 | $13\frac{1}{2}$ | 14 | $16\frac{1}{2}$ | 17 |

   Calculate the mean and standard deviation of these times for each of the two routes. On the basis of these results, which route would you recommend and why? (S.M.P.)

2. The number of live births registered each day in a town during one year has the following frequency distribution:

No. of births	0	1	2	3	4	5	Total
Frequency	60	82	93	72	44	14	365

   Calculate the mean deviation from the mean for this distribution. (J.M.B.)

3. In a factory there are six grades of workmen who earn respectively £700, £800, £850, £950, £1100 and £1150 per year. Equal numbers are employed in each grade. Calculate the standard deviation. (Cam.)

4. The table shows the I.Q. of 100 pupils at a certain school. Calculate (i) the mean, (ii) the mean deviation, (iii) the standard deviation.

I.Q.	55–	65–	75–	85–	95–	105–	115–	125–	135–
No. of pupils	1	3	7	20	32	25	10	1	1

(Note: 55– means "from 55·0 to 64·9 inclusive", each I.Q. being given correct to one decimal place.)   (Cam.)

5. This table shows the shoe sizes of 150 pupils in the secondary schools of a town.

Size	1	$1\frac{1}{2}$	2	$2\frac{1}{2}$	3	$3\frac{1}{2}$	4	$4\frac{1}{2}$	5	$5\frac{1}{2}$	6	$6\frac{1}{2}$	7	$7\frac{1}{2}$	8
Number	2	2	3	5	8	5	7	9	13	11	13	12	9	9	10

Size	$8\frac{1}{2}$	9	$9\frac{1}{2}$	10	$10\frac{1}{2}$	11	$11\frac{1}{2}$	12	$12\frac{1}{2}$	13
Number	9	8	5	4	3	1	0	1	0	1

Form a new table based on sizes 1 and under 3, 3 and under 5, . . ., and use it to calculate the mean and standard deviation. If a small shop concentrates on shoe sizes within $\pm 2S$ of the mean, which of these sizes will not be stocked? Estimate the fraction of these secondary-school pupils who would have to shop elsewhere.   (Cam.)

6. Show that, for any constant a, $\Sigma(x-a)^2 = \Sigma(x-\bar{x})^2 + n(\bar{x}-a)^2$, where the summation extends over all n values of x and $\bar{x}$ is the mean. Hence find the value of a for which $\Sigma(x-a)^2$ is a minimum. Can you suggest a value of a for which $\Sigma|x-a|$ is a minimum?

7. The average age of a class of 20 children is 11 years 3 months and the standard deviation is 5 months. A new pupil whose age is 13 years is added to the class. Calculate the average age and the standard deviation of the new class.   (Cam.)

8. Let S_1^2 denote the variance and $\bar{x}$ the mean of n observations $x_1, x_2, \ldots, x_n$. Suppose that another observation x_{n+1} becomes available, and that the variance of the combined set of $n+1$ observations is S_2^2. Show that

$$S_2^2 = \frac{n}{n+1}S_1^2 + \frac{n}{(n+1)^2}d^2$$

where $d = x_{n+1} - \bar{x}$. What is the least value of S_2 for varying values of x_{n+1}?   (J.M.B.)

9. The table shows the distribution of the number of matches per box in a sample of 100 boxes.

Number of matches	44	45	46	47	48	49	50
Number of boxes	28	28	14	15	8	5	2

Calculate the arithmetic mean and the standard deviation of the number of matches per box in the sample. A second sample of 50 boxes of matches is found to have an arithmetic mean of 47·2 matches per box and a standard deviation of 1·5 matches per box. Calculate the arithmetic mean and the standard deviation of the combined sample of 150 boxes.   (J.M.B.)

10. A set of m observations has mean $\bar{x}$ and variance S_1^2 and a second set of n observations has mean $\bar{y}$ and variance S_2^2. Find the mean of the combined set and show that its variance S^2 is given by:

$$(m+n)S^2 = mS_1^2 + nS_2^2 + \frac{mn}{m+n}(\bar{x}-\bar{y})^2.$$

(Note: question 8 is a special case of this result.)

11. Find the mean and standard deviation of the men's height for the data of Exercise 6A, question 3 (page 70).

12. A sample of 100 deaths registered during a particular period had the following distribution of age, in years:

Age at death:	0–	10–	20–	30–	40–	50–	60–	70–	80–89
Frequency:	14	1	2	6	5	15	17	21	19

Find the mean m and standard deviation S, and evaluate the coefficient of variation V, given by $V = 100S/m$. Give a brief explanation of the double-peaked nature of the above distribution. (J.M.B.)

REVISION EXAMPLES II

1. Pupils on the registers in each age-group in grant-aided and independent schools in January 1958 were as follows:

	Age last birthday	No. of pupils in thousands (nearest 1000)
Nursery	2–4	211
Primary	5–10	4214
Secondary	11–15	2928
Advanced Secondary	16–18	180

Represent the above data as a histogram. (J.M.B.)

2. The following distribution of marks (out of 100) was obtained with a certain examination paper.

Mark range	10–29	30–39	40–49	50–59	60–69	70–79	80–
Frequency	2	8	14	26	28	10	4

Present these marks in a histogram. Calculate the mean mark as well as you can, explaining the limitations of your calculation. Do you consider that the mean or the median is the better measure of average performance for the observations? Give reasons for your answer. (Cam.)

3. Sales by a firm, in units of £100, over 48 consecutive weeks were:

37·2	55·2	93·7	78·6	63·2	77·2	54·3	67·9	60·1	58·0	77·9	67·5
66·1	68·9	49·5	73·8	74·8	80·1	73·1	83·2	66·4	69·7	54·7	68·1
62·5	60·5	55·8	90·1	47·2	71·2	68·2	61·4	40·9	65·1	63·9	93·4
51·0	71·0	71·3	89·1	58·0	46·5	77·2	63·3	53·8	47·4	76·4	95·6

Reduce these results to a frequency distribution with intervals centred at 40, 50, 60, etc. Represent the results by a histogram drawn to scale. On a separate diagram plot the frequencies up to the end of each interval and sketch a cumulative frequency curve. From this curve read off the median, and the quartiles to the nearest £100. (M.E.I.)

4. There are three numbers a, b and c, of which a is the smallest and c the largest. Find for these three numbers:
   (i) the arithmetic mean,
   (ii) the median,
   (iii) the mean deviation from the median,
   (iv) the standard deviation.
   Find an expression for the standard deviation of the four numbers a, b, c and d. (J.M.B.)

5. The marks gained by 100 students taking an examination had the following distribution:

mark	frequency	mark	frequency
0–9	2	50–59	17
10–19	7	60–69	12
20–29	11	70–79	9
30–39	15	80–89	5
40–49	18	90–99	4

Evaluate the mean and standard deviation of this distribution. It is required to assign one of four grades A, B, C, D to each student in such a way that the top 15% of the students receive grade A, the next 35% receive B, the next 35% receive C and the bottom 15% receive D. By drawing the cumulative frequency graph, or otherwise, estimate the lowest mark in each of the grades A, B and C. (J.M.B.)

6. If the frequency of a measurement x is f and x_0 is any number, prove that the mean value M of the measurements is equal to $N^{-1}\Sigma f(x-x_0)+x_0$, where $N = \Sigma f$ and the summations include all non-zero values of f. Prove also that the standard deviation S of the measurements is given by:

$$S^2 = N^{-1}\Sigma f(x-x_0)^2 - (x_0 - M)^2.$$

The frequency f with which x α-particles are emitted from a radioactive specimen in a given time is shown in the following table:

x	0	1	2	3	4	5	6	7	8	9	10	11	12
f	57	203	383	525	532	408	273	139	45	27	10	4	2

Using $x_0 = 4$ as a trial mean, or otherwise, calculate the mean and standard deviation of the observed values of x. (Cam.)

7. The following table gives a frequency distribution of the times, to the nearest second, of the 39 races (heats and final) for the Thames Challenge Cup at the Henley Royal Regatta.

minutes	seconds		minutes	seconds	Frequency
7	10	to	7	19	4
7	20	,,	7	29	8
7	30	,,	7	39	8
7	40	,,	7	49	9
7	50	,,	7	59	6
8	0	,,	8	9	0
8	10	,,	8	19	3
8	20	,,	8	29	1

Plot a cumulative frequency graph and use it to determine:
(i) the 80th percentile,
(ii) the percentile rank of a time of 7 minutes 45 seconds.
Explain the meaning of your result. Calculate the mean time and the standard deviation. (J.M.B.)

8. In an arithmetic examination, five candidates obtained the following marks:

A	B	C	D	E
4	6	8	12	15

Calculate the mean and standard deviation of these marks. It is required to scale these marks so that they shall have a mean of 50 and a standard deviation of 20. Calculate the new values of A, B, C, D, E. (Cam.)

9. The numbers of members, the means and the standard deviations of three distributions are as follows:

No. of members	Mean	s.d.
250	45	8
350	55	10
400	50	9

Calculate the mean and standard deviation of the distribution formed by the three taken together. (J.M.B.)

10. A sample of 76 insects of a certain species was taken and their feeding times were recorded to the nearest 2·5 seconds. Apply a logarithmic transformation to the times given below in order to reduce skewness, and plot a histogram after grouping the transformed observations. Calculate the geometric mean of the feeding times from the grouped data.

Time	50	60	70	75	85	90	95	100	105	110	115	120	125	130	135	140
Frequency	1	1	1	4	2	5	2	12	3	6	5	1	2	1	2	1

Time	145	150	160	165	170	190	200	220	240	250	270	280	340	355	360
Frequency	3	1	3	2	3	2	1	3	·1	1	2	2	1	1	1

(The geometric mean of a sample of values $x_1, x_2, \ldots, x_n$ is defined to be $(x_1 x_2 \ldots x_n)^{1/n}$.) (M.E.I.)

11. Calculate the mean and standard deviation of the following distribution of scores.

Score	1–5	6–10	11–15	16–20	21–25	26–30	31–35
Frequency	3	19	38	69	45	21	5

Obtain the percentage cumulative frequencies and draw the cumulative frequency curve. Find the median and quartile values:

(i) from the curve,

(ii) by interpolation. (J.M.B.)

Discrete Probability Distributions

7.1. A theoretical model: fruit machines

If you look at the frequency charts and histograms which you obtained as a result of the investigations at the beginning of Chapter 4, you will probably notice that the same pattern sometimes crops up in experiments which superficially do not seem to have any connection. For example, the histogram of the estimates of the length of a line and that for the lengths of people's fingers might both look alike. Similarly other shapes occur again and again. Can we forecast what shape the distribution will have before we do the experiment?

- Suppose we start by considering the winnings on a fruit machine. We will take a simplified model in which the three rotating wheels, each providing a display at one of the small windows, are identical and carry ten pictures comprising:

ⓧⓧⓧⓧ 4 oranges ◯◯◯ 3 lemons ♪♪ 2 cherries ♙ 1 bell

The pay-out only occurs for a trio of identical pictures and the winnings in each case are:

ⓧ ⓧ ⓧ 20p, ◯◯◯ 30p, ♪♪♪ 50p, ♙♙♙ 100p.

What distribution of winnings would we expect to get, assuming that every picture on each wheel has an equal chance of appearing and that the three wheels are independent? And what would we expect the mean pay-out per round to be?

We can compute the probabilities of each result directly using the multiplication law. Let $P(x)$ stand for the probability that the winnings are x pence. Then

$P(20) = P(ⓧⓧⓧ) = P(ⓧ).P(ⓧ).P(ⓧ)$ since the three wheels are taken to be independent,

$= \frac{4}{10} \cdot \frac{4}{10} \cdot \frac{4}{10} = 0 \cdot 064$

similarly $P(30) = P(\bigcirc\bigcirc\bigcirc) = \frac{3}{10}\cdot\frac{3}{10}\cdot\frac{3}{10} = 0\cdot027$

$P(50) = P(\;\;) = \frac{2}{10}\cdot\frac{2}{10}\cdot\frac{2}{10} = 0\cdot008$

$P(100) = P(\;\;) = \frac{1}{10}\cdot\frac{1}{10}\cdot\frac{1}{10} = 0\cdot001$

$P(0) = P(\text{any other result}) = 1 - (0\cdot064 + 0\cdot027 + 0\cdot008 + 0\cdot001) = 0\cdot9.$

So if we suppose that we played the machine 1000 times we would expect the frequencies to be something like 64, 27, 8, 1, 900 and the frequency chart is shown in Fig. 7.1.

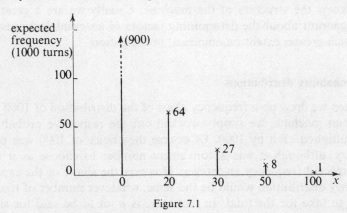

Figure 7.1

Of course it is highly unlikely that the results of 1000 actual turns (assuming we had both the money and the courage) would give exactly this picture, just as it is highly unlikely that in 1000 tosses of a fair coin the number of tails would be exactly 500; but in both cases we would expect the actual results to be "reasonably" close to the predicted ones.

What would we expect the average pay-out per round to be?

Assuming the results for the 1000 turns are exactly as predicted, then, if the mean winnings are $\bar{x}$ pence we have

$\bar{x} = (64 \times 20 + 27 \times 30 + 8 \times 50 + 1 \times 100)/1000 = 2590/1000 = 2\cdot59.$

Hence the mean winnings per turn would be 2·59p.

If the machine required a stake of 3p per turn, then we would expect to lose on average 0·41p per turn and in 1000 turns we would expect to lose £4·10.

What we have built up in the above investigation is a theoretical "model" of the actual situation. The whole inquiry began with the question, "What are our expected winnings?" In order to answer this we had to construct a model, making on the way various assumptions (that the wheels were independent and that the appearance of each picture was equally likely).

Having set up a probability model, we proceeded to work out the expected distribution, and hence the expected mean winnings per turn.

In practice the next step would be to perform an experiment of as many turns on the fruit machine as possible. The distribution of these would then be plotted and compared with the expected result. The assumptions could then be either accepted or modified according as the two distributions were similar or different.

Note that in this particular case our assumptions were easy to make, as we knew the structure of the machine. Usually we are a great deal more ignorant about the determining factors of an event and must rely to a much greater extent on empirical investigation.

7.2. Probability distributions

When we drew up a frequency chart of the distribution of 1000 turns on a fruit machine, we simply worked out the respective probabilities and multiplied each by 1000. Of course the choice of 1000 was purely arbitrary, although it was a convenient number to choose as it made each expected frequency an integer. However the shape of the expected frequency distribution would be the same, whatever number of trials we agreed to take for the total. In fact there is a lot to be said for simply drawing the chart showing the probabilities. The conversion to any agreed total can then be simply made and the chart will have the virtue that the sum of lengths of the columns will now be 1. This form of the distribution is known as a **probability distribution.** It is obviously analogous to a *frequency distribution,* the difference being that a probability distribution is simply an outcome space with a probability value assigned to each outcome, whereas the frequency distribution is experimentally determined, and assigns to each outcome the number of times it has actually occurred. A more exact parallel can be drawn if, in the experimental case, we draw a chart showing the *relative frequency f/n* in each case; for then the total sum will be 1 as in the theoretical case.

A consequence of doing this is that the result will be independent of the number of trials n. However we will not in general use a relative frequency distribution because a knowledge of n is necessary for the evaluation of the significance of the data. Also, we do not wish to obscure the very real differences between a frequency distribution based on an actual experiment, and a probability distribution calculated on the basis of a model we have assumed to fit the circumstances.

In general then we have the two cases shown in Fig. 7.2 and Fig. 7.3. We call the function $P(x)$ the **probability function.**

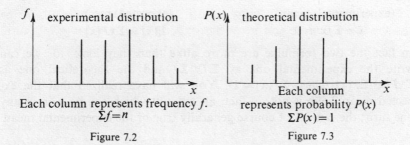

Each column represents frequency f.
$\Sigma f = n$

Figure 7.2

Each column represents probability $P(x)$
$\Sigma P(x) = 1$

Figure 7.3

If we return to our fruit machine, we worked out that the mean "winnings" per turn (where by "winnings" we mean the actual sum paid back by the machine) was 2·59p.

We found this mean in the following way:

winnings (x)	probability function $P(x)$	expected frequency/ 1000 turns (f)	fx
100	0·001	1	100
50	0·008	8	400
30	0·027	27	810
20	0·064	64	1280
0	0·9	900	0
	$\Sigma P(x) = 1$	$n = \Sigma f = 1000$	$\Sigma fx = 2590$

mean $= \Sigma fx/n = 2590/1000 = 2\cdot59$

However if we examine this we see that introducing an ideal total of 1000 was quite unnecessary; after all, the mean winnings per turn should be the same whatever we take for n. In fact if we look hard enough at the working, we will notice that we have first multiplied all the $P(x)$'s by 1000 (to get f) and then in the last line we have divided the final total by 1000 to get the mean. We might just as well have saved ourselves the trouble and calculated the mean as $\Sigma x P(x)$. In this particular case we can check our new formula as

$\Sigma x P(x) = 100 \times 0\cdot001 + 50 \times 0\cdot008 + 30 \times 0\cdot027 + 20 \times 0\cdot064$
$= 0\cdot10 + 0\cdot40 + 0\cdot81 + 1\cdot28 = 2\cdot59.$

The result is exactly the same.

We have stopped referring to the mean as $\bar{x}$. It is usual to use $\bar{x}$ for the mean of a set of experimental data calculated as $\Sigma fx/n$ and to write $E(x)$ or sometimes the Greek letter μ (written "mu" and pronounced "mew") for the mean of a theoretical distribution calculated from $\Sigma x P(x)$. We shall sometimes call $E(x)$ the **expected mean** or **expectation,** though we shall still just call it the mean where there is no risk of confusion. We can summarize these results as:

(experimental) mean	(expected) mean
$\bar{x} = \Sigma fx/n$	$E(x) = \Sigma x P(x)$

In fact the two formulae are more alike than they look, for we can write the experimental one as $\Sigma fx/\Sigma f$ and the theoretical one as $\Sigma x P(x)/\Sigma P(x)$ (since $\Sigma P(x) = 1$). You will have realized that the expected mean 2·59p is not, in fact, an amount which can be won in any one turn; the same is of course generally true of the experimental mean.

7.3. Variance and standard deviation

It will come as no surprise that, just as we find the expected mean from a probability distribution, so we can find the expected standard deviation. In practice the (expected) variance, the square of the standard deviation, is used more often than the standard deviation itself. As before we will only insert the word "expected" when there is any fear of confusion.

In the case of the experimental frequency distribution, we defined the variance S^2 to be the mean squared deviation from the mean, and this led us to the formula

$$S^2 = \frac{\Sigma f(x - \bar{x})^2}{n}$$

The corresponding definition of the expected variance $\mathrm{Var}(x)$ is the expected mean squared deviation from the expected mean which we can write as $\mathrm{Var}(x) = E[(x - E(x))^2]$. This is also called σ^2 (pronounced sigma-squared). By analogy with $E(x) = \Sigma x P(x)$ we have

$$\mathrm{Var}(x) = E[(x - E(x))^2] = \Sigma(x - E(x))^2 P(x).$$

Just as in the experimental case we found $S^2 = \frac{\Sigma f(x - \bar{x})^2}{n}$ simplified to $S^2 = \frac{\Sigma fx^2}{n} - \bar{x}^2$, so in the theoretical case we can prove the alternative formula $\mathrm{Var}(x) = \Sigma x^2 P(x) - (E(x))^2 = E(x^2) - (E(x))^2$. This latter version is the most useful for calculation.

We now have:

(experimental) variance	(expected) variance
$S^2 = \dfrac{\Sigma f(x - \bar{x})^2}{n}$	$\mathrm{Var}(x) = \Sigma(x - E(x))^2 P(x)$
$= \dfrac{\Sigma fx^2}{n} - \bar{x}^2$	$= \Sigma x^2 P(x) - (E(x))^2.$

We can work out the variance for the fruit machine by adding an extra column to the previous table.

x	$P(x)$	$xP(x)$	$x^2P(x)$
100	0·001	0·1	10
50	0·008	0·4	20
30	0·027	0·81	24·3
20	0·064	1·28	25·6
0	0·9	0	0

$$\Sigma P(x) = 1 \quad \Sigma xP(x) = 2·59 \quad \Sigma x^2P(x) = 79·9$$

$\text{Var}(x) = 79·9 - (2·59)^2 \simeq 73·19$. Standard deviation $\sigma \simeq \sqrt{73·19} \simeq 8·6$.

In other words if we played with the fruit machine for long enough and kept a record of our winnings, we would expect the mean winnings per turn to be 2·59p and the standard deviation to be roughly 8·6p.

7.4. A worked example: score on a die

To illustrate the use of the above formulae in a rather simpler situation let us suppose that we wanted to work out what the mean and variance would be if we kept on throwing an unbiased die for long enough.

In fact the mean can be obtained without using a formula if we remember that the set of possible scores is $\{1, 2, 3, 4, 5, 6\}$ and all the scores are equally likely. However we can check it mathematically and also obtain the variance, the value of which is rather less easy to guess.

If we look at the probability distribution, we find all is very simple as each probability is exactly $\frac{1}{6}$.

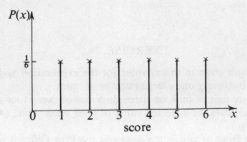

Figure 7.4

This sort of distribution is known as a *rectangular* or *uniform* distribution for obvious reasons.

$$E(x) = \Sigma xP(x) = \frac{1}{6}.1 + \frac{1}{6}.2 + \frac{1}{6}.3 + \frac{1}{6}.4 + \frac{1}{6}.5 + \frac{1}{6}.6$$
$$= \frac{1}{6}.21$$
$$= 3·5$$

So the mean is, very reasonably, 3·5.

Similarly $\text{Var}(x) = \Sigma x^2 P(x) - (E(x))^2$
$$= \tfrac{1}{6}.1^2 + \tfrac{1}{6}.2^2 + \tfrac{1}{6}.3^2 + \tfrac{1}{6}.4^2 + \tfrac{1}{6}.5^2 + \tfrac{1}{6}.6^2 - (3\cdot5)^2$$
$$= \tfrac{91}{6} - \tfrac{49}{4} = \tfrac{35}{12}.$$
So the variance is $\tfrac{35}{12} \simeq 2\cdot92$.

7.5. Random variables

In both these examples we were considering the expected mean of a quantity; in the first it was the winnings on a turn at a fruit machine, and in the second it was the score of a die. Both these quantities are variables in that they can take many possible values. Because, in performing an actual experiment, we would be unable to forecast exactly which value of the variable would turn up at the next trial, they are both said to be **random variables.**

In fact the whole of probability and statistics is concerned with random variables. The number of heat quanta per atom, the height of a man, and the length of time spent waiting for a bus are all examples; the exact value at the next instance encountered is not determinable in advance, but we can estimate the probability of each variable taking any given value or range of values.

When in future we refer to a random variable x, we simply mean a measurable quantity which can take any of a number or range of values, and does so with a given probability distribution. (In later work it is often necessary to distinguish between the actual random variable denoted by X and the values it takes x. Generally we will use x for both.)

EXERCISE 7A

1. Check the values given in this chapter for the expectation and variance of the score of a die by tossing one a large number of times.
2. If you have one pint of milk on three days out of every four and none on the fourth day, calculate the expected value of your weekly bill. (Assume a pint of milk costs 10p.)
3. The person in front of you in the queue at the Post Office is sure to be either:
    (i) sending a parcel to China, which takes 10 minutes,
    (ii) opening a Giro account, which takes 8 minutes, or
    (iii) sending a telegram to John O'Groats, which takes 5 minutes.
   If the probabilities of these activities are $\tfrac{1}{2}$, $\tfrac{1}{6}$ and $\tfrac{1}{3}$ respectively, find the mean and variance of the time you have to wait for him. If you are always buying stamps, which takes 1 minute, find the mean and variance of the time the next person in the queue has to wait for both of you.
4. Using the results of the previous question, write down expressions for $E(x+a)$ and $\text{Var}(x+a)$, where a is any constant, in terms of $E(x)$, $\text{Var}(x)$ and a.

5. In your pocket you have three 1p pieces, two 2p pieces, two 5p pieces and one 10p piece. You take two coins out at random and put them in a collecting box. What is the expected amount that you give?

6. 90% of first-class mail arrives the day after being posted, the other 10% takes one day longer; for second-class mail the corresponding figures are 20% and 40%, while a further 30% takes three days, and the remainder four days. If two letters out of every five go by first-class mail, what is the mean delivery time for letters?

7. A farmer has 24 hurdles, each 2 m long, with which to make a rectangular sheep-pen. If he uses all the hurdles and is equally likely to choose any number for the side parallel to the road, find the mean and standard deviation of the area of the pen.

8. Bill earns £20 for a five-day week. He works every fourth Sunday, for which he is paid double time, and every alternate Saturday, for which he earns time and a half. Draw up a table showing the probability distribution of his wage for a day chosen at random, and hence find the mean and variance of his daily wage. Find also the mean and possible variances of his weekly wage. (Why are there two possible values for the variance?)

9. A game consists of tossing three unbiased pennies. If the number of £s won on any play of the game is equal to the square of the number of heads showing, what are the average winnings per game? If the winnings in each case now become three times what they were before, write down the new average winnings. Can you use this result to write down an expression for $E(ax)$, where a is any constant, in terms of $E(x)$ and a? What is the corresponding result for the variance?

10. Two unbiased dice are thrown together. Tabulate the probability distribution of the sum of the two scores. If a player wins 20p for a score of 12, and 10p for a score of 10 or 11, but has to pay 2·5p for each round, calculate his expected loss in a hundred rounds.

11. Calculate the mean and variance of the sum of the scores on two dice.

12. In theory random digits are uniformly distributed. Calculate the expected mean and variance of the entries in a table of random numbers and check them (if you have not done so already) by reference to the table at the back of the book.

13. If two kilometre posts are selected at random on a stretch of road n kilometres long (starting with a post), show that their average distance apart will be $(n+2)/3$ km.

14. A die is designed so that the probability of its landing on any side is proportional to the number on that side. Calculate the probability of its landing on each side and hence the mean and variance of the score of the die.

15. In a pocket game of cricket two hexagonal cross-sectioned metal pieces are rolled together. One piece is labelled with the scores 1, 2, 3, 4, 6, 4 and the other is labelled 1, 2, 3, 1, OUT, OUT. Find the distribution of the total score from one throw of the two pieces. (OUT rules out the score on the other piece, e.g. 2, OUT means a score of zero.) State the probability of:

(a) getting out, (b) scoring 5 or more, (c) scoring exactly 6.

Find the probability that a player survives two throws of the pieces and goes out on the third.

(S.M.P.)

16. In a certain game a player has three throws of an unbiased die. At each throw he scores according to the face uppermost, if it is a six or if he has previously thrown a six, but not otherwise. Tabulate the probability distribution of his total score. Show that his chance of scoring more than 11 is less than 1 in 6, but that his chance of scoring more than 10 exceeds 1 in 6. (Cam.)

17. In order to use the labels in a competition, a housewife removes the labels from 3 tins of tomato soup and 4 tins of peaches, but forgets to mark the tins which, without their labels, are all identical. The housewife opens successive tins, chosen at random, looking for a tin of peaches. If p_r is the probability that the first tin of peaches is opened at the rth attempt, calculate p_r for $r = 1, 2, 3, 4$. Calculate also the expectation and the variance of the number of attempts needed. (Cam.)

18. Two men play a game with two dice. A has a true die, whereas B has a die which is biased so that each of the even faces is twice as likely to occur as each of the odd faces. The two players throw their own dice, and A wins from B the sum of the numbers thrown when the sum is even and the numbers are unequal. B wins the sum from A when it is odd; and the game is drawn when the two numbers are equal. Calculate the expectation of the game to A.

(M.E.I.)

19. In a television panel game each player is asked a series of questions until either he answers one incorrectly or he has answered three correctly; his score for the turn is the number of correct answers he has given, with a bonus mark if he gives 3 correct answers. Assume that the questions are independent, and that the chance of giving a correct answer to any question is p. Obtain the mean score for a turn. Show that for $p = \frac{1}{2}$ the chance that a team of three players obtain a total score over 3 after one turn each is $\frac{25}{64}$. (Cam.)

20. What is the expected number of moves that can be made by:
   (i) a bishop, (ii) a knight, placed at random on an empty chess board?

7.6. The cumulative distribution function

We first decided in Chapter 5 that we were interested in the number of holes in a game of golf in which the score was less than or equal to a given number, and so we came to define the cumulative frequency table.

In the same way we might well want to know how likely we are to win 30p or less on a single turn at the fruit machine, or how likely we are to throw a number less than or equal to 3 with the die.

What we need in the first case is $P(x \leqslant 30)$ where x denotes the winnings in one turn on the machine. We can easily obtain this from the probability distribution:

x	$P(x)$
0	0·900
20	0·064
30	0·027
50	0·008
100	0·001

$P(x \leqslant 30) = P(0) + P(20) + P(30) = 0.900 + 0.064 + 0.027 = 0.991$, since if x is to be no greater than 30, it must take one of the three values 0, 20 or 30. We could derive $P(x \leqslant 0)$, $P(x \leqslant 20)$, $P(x \leqslant 50)$, $P(x \leqslant 100)$ in the same way and obtain a new function, which we shall call the *cumulative distribution function* (c.d.f.) and denote by F as we did in the experimental case. In this example we have:

x	$F(x)$
0	0.900
20	0.964
30	0.991
50	0.999
100	1.000

We can also give the values of F for intermediate values of x, for example we have $F(25) = P(x \leqslant 25) = P(x \leqslant 20) = 0.964$, so you can see that if we drew a graph of $F(x)$ for all values of x it would look like Fig. 7.5.

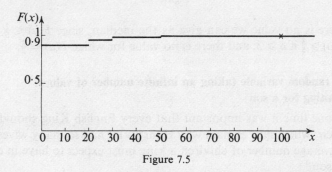

Figure 7.5

Here again we have a step function. You will notice that if a is less than the smallest value taken by x then $F(a) = 0$, and if a is greater than or equal to the largest value taken by x then $F(a) = 1$. Also you will notice that F is always increasing. (If x can take an infinite number of values, then we may not be able to talk about the largest and smallest values of x, and we can only say that $F(a) \to 1$ as a gets large and positive and $F(a) \to 0$ as a gets large and negative.) Notice that we could write down a formula for $F(a)$:

$$F(a) = \sum_{x \leqslant a} P(x)$$

where the summation is taken over all values of x less than or equal to a.

There is in theory no reason why we should not define the median and quartiles of the theoretical distribution as we did in the experimental case, as being the values a for which $F(a) = \frac{1}{2}$ or $F(a) = \frac{1}{4}$ or $F(a) = \frac{3}{4}$ respectively, and this is often done. The snag is that there may not be any value of a for which $F(a) = \frac{1}{2}$, say, and we cannot get round the difficulty as in the experimental case by taking the mean of two values which are close to it. For example suppose a variable x takes each of the values 1 to 5 with probability $\frac{1}{5}$; then we can check that its cumulative distribution function looks like Fig. 7.6.

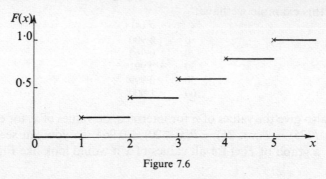

Figure 7.6

There is no value we can give as the median, since $F(a) \leqslant \frac{2}{5}$ if $a < 3$ and $F(a) \geqslant \frac{3}{5}$ if $a \geqslant 3$, and there is no value for which $F(a) = \frac{1}{2}$.

7.7. A random variable taking an infinite number of values: waiting for a son

At one time it was important that every English King should have a male heir, which may explain why Henry VIII had so many wives! What is the average number of children a king must expect to have in order to have a son?

If we look carefully at the question we can see that there is a very small probability that he will get 20 girls before he has a boy, and an even smaller but definitely non-zero probability of his getting 200 girls first. In fact the number of girls coming first could be any whole number, so that we have infinitely many possibilities to consider.

Suppose we assume that the probability of the child's being a boy is $\frac{1}{2}$. We are now definitely making a simplified mathematical model of the situation, as there are in fact slightly more boys than girls born (though this may not have been true in Tudor times). We will also have to assume that the sex of any one child is independent of the sex of all children born before, which luckily accords fairly well with the experimental evidence.

$P(\text{first child is male}) = \frac{1}{2}$ and so $P(1) = \frac{1}{2}$

$P(\text{second child is first boy}) = P(\text{first child is a girl and second child}$
$\text{is a boy})$

$= P$(first child is a girl). P(second child is a boy)

(since we are assuming the second child independent of the first)

$$= \tfrac{1}{2} \cdot \tfrac{1}{2} = \tfrac{1}{4}$$

So $P(2) = (\tfrac{1}{2})^2 = \tfrac{1}{4}$

Similarly

$P(x) = P$(first $x-1$ children are girls and xth child a boy)
$$= (\tfrac{1}{2})^{x-1} \cdot \tfrac{1}{2} = (\tfrac{1}{2})^x$$

So the probability function looks like Fig. 7.7.

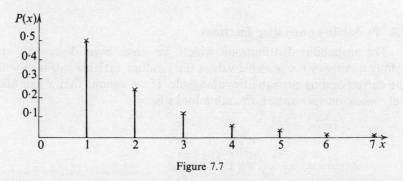

Figure 7.7

We can easily check that the sum of the terms is 1 by directly summing them, but Fig. 7.8 should make it obvious without this trouble.

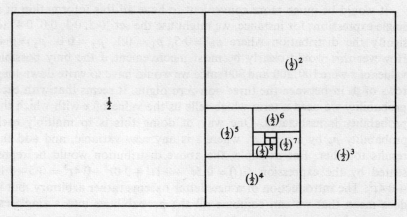

Figure 7.8

What we really wanted to know was the expected number of children required to give the king a male heir.

We have
$$E(x) = \Sigma x P(x) = \Sigma x(\tfrac{1}{2})^x = 1.\tfrac{1}{2} + 2.(\tfrac{1}{2})^2 + 3.(\tfrac{1}{2})^3 + 4.(\tfrac{1}{2})^4 + \ldots = T, \text{ say.}$$
We can evaluate this as follows:
(multiplying both sides by $\tfrac{1}{2}$ we get) $1.(\tfrac{1}{2})^2 + 2.(\tfrac{1}{2})^3 + 3.(\tfrac{1}{2})^4 + \ldots = \tfrac{1}{2}T$
Subtracting the two equations $1.(\tfrac{1}{2}) + 1.(\tfrac{1}{2})^2 + 1.(\tfrac{1}{2})^3 + \ldots = \tfrac{1}{2}T$
But we know the left-hand side is equal to 1 and therefore $T = 2$.
So on average the king must have 2 children in order to get a boy.

The method of evaluating the mean was rather cumbersome and it would be even worse to evaluate the variance in the same way. We will introduce a simpler method of obtaining both in the next section.

7.8. Probability generating functions

The probability distributions which we have been discussing are simply a table of the possible values the random variable can take with the corresponding probabilities alongside. If we assume that x can take only *whole-number values,* the table looks like:

x	$P(x)$
0	p_0
1	p_1
2	p_2
3	p_3
$\vdots$	$\vdots$

(where now we are using p_0 to represent $P(0)$. etc.)

It would be much more convenient to have all this information in a single expression; for instance, we might use the set $\{0.5, 0.1, 0.0, 0.4\}$ to signify the distribution where $p_0 = 0.5$, $p_1 = 0.1$, $p_2 = 0.0$, $p_3 = 0.4$. However this would clearly be most inconvenient if the only possible values of x were 100, 200 and 300 since we would have to write down long rows of 0s in between the three non-zero digits. It seems that with each probability we need a term which tells us the value of x with which the probability is associated. One way of doing this is to multiply each probability p_x by a term t^x, where t is any new variable, and add the results together. For example, the above distribution would be represented by the expression $G(t) = 0.5t^0 + 0.1t^1 + 0.0t^2 + 0.4t^3 = 0.5 + 0.1t + 0.4t^3$. The introduction of a new letter t seems rather arbitrary, but it does mean that we can combine all the probabilities into a single expression which is a polynomial in t (an expression only involving powers of t). We can also tell at a glance which probability fits each value of x, as we only have to look at the coefficient of that power of t. For example, if we want p_3 in the above, we just look at the term involving t^3 and see that $p_3 = 0.4$. Similarly $p_2 = p_4 = 0$ as there are no terms involving t^2 or t^4.

To take another example, if we go back to the die, then we have $p_x = \frac{1}{6}$ for each value of x from 1 to 6 and $p_x = 0$ otherwise. Thus the expression $G(t)$ becomes

$$G(t) = \tfrac{1}{6}t^1 + \tfrac{1}{6}t^2 + \tfrac{1}{6}t^3 + \tfrac{1}{6}t^4 + \tfrac{1}{6}t^5 + \tfrac{1}{6}t^6 = \tfrac{1}{6}(t + t^2 + t^3 + t^4 + t^5 + t^6).$$

You can probably guess what distribution has a corresponding expression $\frac{1}{2} + \frac{1}{2}t$.

In general we will have an expression $G(t) = p_0 + p_1 t + p_2 t^2 + p_3 t^3 + \dots$ which we can write as $G(t) = \Sigma t^x p_x$ where the summation is taken over all the possible values of x.

This looks rather like some formulae which we have had before: $E(x) = \Sigma x P(x)$, which we can write as $\Sigma x p_x$ if x takes whole-number values only, and also $E(x^2) = \Sigma x^2 P(x) = \Sigma x^2 p_x$.

If we compare these with $\Sigma t^x p_x$ by analogy, we can write this as $E(t^x)$. So that $G(t) = E(t^x)$.

We still have not really got much further except that we have given a formula for our expression. As well as being the expected value of t^x, where x takes each value in the range with probability p_x, we usually call the expression the **probability generating function** (p.g.f.).

If we look again at the function $G(t)$ and find what happens when we put $t = 1$, we have $G(1) = p_0 + p_1 + p_2 + p_3 + \dots = 1$ since this is simply the sum of the probabilities associated with x. We could also have seen this since $G(1) = E(1^x) = E(1) = 1$.

Though this does not tell us anything new, it does give us one of the conditions for any polynomial to be a probability generating function and hence give rise to a possible probability distribution.

Can we get from $G(t)$ an expression for the mean,

$$E(x) = \Sigma x p_x = 0 \cdot p_0 + 1 \cdot p_1 + 2 \cdot p_2 + \dots ?$$

In fact the answer is yes, for if we differentiate $G(t)$ with respect to t we get:

$$\frac{dG}{dt} = \frac{d}{dt}(p_0 + p_1 t + p_2 t^2 + p_3 t^3 + \dots) = p_1 + 2p_2 t + 3p_3 t^2 + 4p_4 t^3 + \dots$$

and if we put $t = 1$, $G'(1) = \left|\frac{dG}{dt}\right|_{t=1} = p_1 + 2p_2 + 3p_3 + \dots = E(x)$.

(In future we will always write $G'(t)$ to stand for $\dfrac{dG}{dt}$ so that $\left|\dfrac{dG}{dt}\right|_{t=1}$ can be written as $G'(1)$, which is the result of differentiating and then putting $t = 1$.)

Suppose we try differentiating again:
$$G''(t) = 2p_2 + 2.3p_3 t + 3.4p_4 t^2 + 4.5p_5 t^3 + \ldots$$
so $\;G''(1) = 2p_2 + 2.3p_3 + 3.4p_4 + 4.5p_5 + \ldots = \Sigma x(x-1)p_x$
$$= E[x(x-1)].$$

We can write $\Sigma x(x-1)p_x$ as $\Sigma x^2 p_x - \Sigma x p_x = E(x^2) - E(x)$.
So we have the result
$$G''(1) = E(x^2) - E(x) = E(x^2) - G'(1).$$
so $\;\mathrm{Var}(x) = E(x^2) - [E(x)]^2 = G''(1) + G'(1) - [G'(1)]^2.$

Let us see how to use these results in practice by returning to the problem of the king's son. Here our variable took whole-number values and we had $p_x = (\frac{1}{2})^x$ for $x \geqslant 1$. This gives us the probability generating function $G(t)$.
$$G(t) = p_0 + p_1 t + p_2 t^2 + p_3 t^3 + \ldots = 0 + \tfrac{1}{2}t + (\tfrac{1}{2})^2 t^2 + (\tfrac{1}{2})^3 t^3 + \ldots$$
$$= (\tfrac{1}{2}t) + (\tfrac{1}{2}t)^2 + (\tfrac{1}{2}t)^3 + \ldots.$$

But this series is a geometric progression with first term $\frac{1}{2}t$ and common ratio $\frac{1}{2}t$ so that $G(t) = \dfrac{\frac{1}{2}t}{1 - \frac{1}{2}t} = \dfrac{t}{2-t}$, provided $|t| < 2$.

This is where the power of the probability generating function comes in, as we have a simple formula $G(t) = \dfrac{t}{2-t}$ instead of an infinite sequence of probabilities. Note that all is well as $G(1) = \dfrac{1}{2-1} = 1$.

Now we have found $G(t)$ there is little trouble in finding the mean as
$$G'(t) = \frac{(2-t) - t(-1)}{(2-t)^2} = \frac{2}{(2-t)^2} \text{ so } G'(1) = 2, \text{ confirming our earlier result.}$$
We can also find the variance for
$$G''(t) = \frac{+4}{(2-t)^3}$$
so $G''(1) = 4$ and $\mathrm{Var}(x) = G''(1) + G'(1) - [G'(1)]^2 = 4 + 2 - 4 = 2.$

EXERCISE 7B

1. A die is tossed until a 6 is obtained. Find the probability function of the number of throws required. Find also:
   (i) the most probable number of tosses,
   (ii) the mean number of tosses.
   Tabulate the distribution function and find the least number of tosses n such that there is a greater-than-even chance that a 6 will have been thrown by the nth toss.
2. In the so-called St Petersburg game the player throws a coin until a head is obtained and then receives from the bank £2^n, where n is the number of throws. What is the chance that he receives (a) £16, (b) more than £32? What is the

chance that, if the bank holds £10^6, the player will break the bank in a single play? What can be said about the player's average receipt if the bank's resources are unlimited? (Cam.)

3. A coin has probability p of falling heads and probability $q = 1-p$, of falling tails. The coin is thrown (separate throws being independent) until two consecutive heads occur for the first time. Let p_n denote the probability that two consecutive heads occur for the first time on the nth and $(n+1)$th throws. Calculate p_1 and p_2. Show, by considering the outcomes of the first two throws, that $p_n = qp_{n-1} + pqp_{n-2}$, for $n \geqslant 3$. By multiplying both sides of this equation by n, and summing, show that the mean number of throws needed to obtain two consecutive heads for the first time is $(1+p)/p^2$. (J.M.B.)

4. A random variable has p.g.f. $\frac{1}{6} + \frac{1}{3}t + \frac{1}{4}t^3 + \frac{1}{6}t^4 + \frac{1}{12}t^5$. Tabulate its c.d.f. and find its mean and variance.

5. A random variable has probability generating function $(1+2t)^4/81$, find $P(2)$, $P(4)$, μ and σ^2.

6. A random variable has p.g.f. $a + bt + ct^2$. It has mean $\frac{7}{6}$ and variance $\frac{29}{36}$. Find the values of a, b and c.

7. A learner driver has constant probability $\frac{1}{3}$ of passing the test at any attempt. Find the p.g.f. of the number of tests he takes before passing. Hence find the mean and variance of this number of tests.

8. The random variable Y has a distribution such that $P(Y = r)$ is given by the coefficient of θ^{t-r} in $p^t(1-\theta q)^{-t}$ for $r \geqslant t$ and zero otherwise, where $0 < p < 1$, $q = 1-p$. Find the expectation of Y. (M.E.I.)

9. The last ball in a game of bar billiards must be hit down a specified hole by one of the two players, A or B. If A has the first shot and his probability of holing the ball is $\frac{1}{6}$ and B's probability is $\frac{1}{4}$, find the probability that A holes the ball, and also the expected number of shots before the ball is holed.

10. Questions 1, 2, 7 and the 'King's son' problem are particular instances where a suitable model is provided by the *geometric distribution* which has
$$p(x) = (1-p)^{x-1}p \qquad x = 1, 2, 3 \ldots, 0 < p < 1.$$
Show that
$$E(x) = \frac{1}{p}, \qquad \text{Var}(x) = \frac{1-p}{p^2}$$
and
$$G(t) = \frac{pt}{1-t+pt}, \text{ provided } |t| < 1/(1-p).$$
Can you think of other situations where the model might be used?

The Binomial and Poisson Distributions

8.1. At the shooting range

George and his friend Jim are at a fairground, trying their skill at the shooting range. In a round they each have five shots at a ping-pong ball balanced on a jet of water. After each successful shot the ball is replaced. They have been playing long enough to estimate their skill, and they reckon that the probability of a success on any shot is $\frac{1}{2}$ for Jim and $\frac{1}{3}$ for George, and that these figures are unaffected by the number of previous hits in the round. George, being statistically minded, wants to work out their respective chances of winning a prize in any round (which means scoring at least four out of the possible five hits), and also the average number of hits they will each make in a round.

How does he set about this problem? The random variable he is interested in is the number of hits in a round, x say, where x can take any integer value from 0 to 5. He needs then to find the distribution of x. Let us consider Jim's case first and argue as follows:

$P(0) = P(\text{MMMMM})$ where M stands for the shot being a miss and MMMMM for the event that all five shots in the round are misses. Given that each shot is independent of all the previous shots in the round, we can say that $P(\text{MMMMM}) = [P(\text{M})]^5 = (\frac{1}{2})^5$ since the probability of Jim scoring a hit on any shot is $\frac{1}{2}$ and therefore the probability of a miss is also $\frac{1}{2}$.

Hence $P(0) = (\frac{1}{2})^5 = \frac{1}{32}$.

So far so good, but what about $P(1)$? We can see that this will be given by

$P(\text{HMMMM} \cup \text{MHMMM} \cup \text{MMHMM} \cup \text{MMMHM} \cup \text{MMMMH})$,

where HMMMM stands for the event "a hit on the first shot and then four misses", etc. These events are disjoint and therefore

$P(1) = P(\text{HMMMM}) + P(\text{MHMMM}) + P(\text{MMHMM}) + P(\text{MMMHM}) + P(\text{MMMMH})$.

100

Reasoning as before $P(\text{HMMMM}) = P(\text{H}).[P(\text{M})]^4 = \frac{1}{2}.(\frac{1}{2})^4 = (\frac{1}{2})^5$; the same is true for all the other terms and so we can say $P(1) = 5.(\frac{1}{2})^5 = \frac{5}{32}$.

Can we find $P(2)$ by a similar method without writing down every possibility? The event $x = 2$ will be the union of events like HHMMM, where each event consists of 2 hits and 3 misses. There will be $\binom{5}{2}$ of these events, which is the number of ways of choosing the 2 shots which are to be hits, and each will have the same probability $[P(\text{H})]^2 [P(\text{M})]^3$ $= (\frac{1}{2})^5$. Therefore $P(2) = \binom{5}{2}(\frac{1}{2})^5 = \frac{10}{32}$.

Arguing similarly, $P(3) = \binom{5}{3}(\frac{1}{2})^5 = \frac{10}{32}$,
$$P(4) = \binom{5}{4}(\frac{1}{2})^5 = \frac{5}{32},$$
$$P(5) = \binom{5}{5}(\frac{1}{2})^5 = \frac{1}{32}.$$

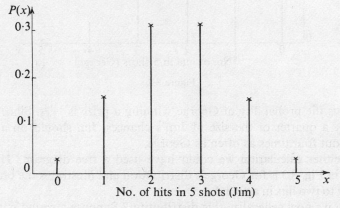

Figure 8.1

We can write out Jim's probability function, which we can see is symmetric, as:

x	0	1	2	3	4	5
$P(x)$	$\frac{1}{32}$	$\frac{5}{32}$	$\frac{10}{32}$	$\frac{10}{32}$	$\frac{5}{32}$	$\frac{1}{32}$

As we expect, we see that $\Sigma P(x) = 1$.

We can now say that the probability of Jim winning a prize, which is equal to $P(4) + P(5)$, is $\frac{3}{16}$.

What will happen in George's case? The only difference here is that $P(\text{H}) = \frac{1}{3}$ and $P(\text{M}) = \frac{2}{3}$, so $P(\text{H}) \neq P(\text{M})$. Otherwise we can argue as before. For instance, $P(0) = P(\text{MMMMM}) = [P(\text{M})]^5 = (\frac{2}{3})^5$.

$P(1)$ is still the sum of terms like $P(\text{HMMMM})$. Each of these component probabilities is still $P(\text{H}).[P(\text{M})]^4$, which now equals $\frac{1}{3}(\frac{2}{3})^4$. Thus $P(1) = \binom{5}{1}.\frac{1}{3}.(\frac{2}{3})^4 = \frac{80}{243}$.

Working similarly we arrive at $P(2)$, $P(3)$, $P(4)$ and $P(5)$ and obtain George's probability table:

x	0	1	2	3	4	5
$P(x)$	$\frac{32}{243}$	$\frac{80}{243}$	$\frac{80}{243}$	$\frac{40}{243}$	$\frac{10}{243}$	$\frac{1}{243}$

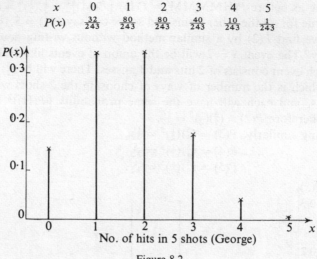

No. of hits in 5 shots (George)

Figure 8.2

Thus the probability of George winning a prize is $\frac{11}{243}$. Since this is roughly a quarter of the size of Jim's chances, Jim should on average win about four times as often as George.

In either calculation we could have used a tree diagram. The one shown in Fig. 8.3 is for George's distribution and illustrates the branches leading to two hits in a round.

Can we now generalize this distribution? Suppose a round consisted of n shots and the probability of a hit on any one of these was p. Again let x be the number of hits in one round; x can then take any value between 0 and n. Can we find a formula for $P(x)$?

The event "x hits" will be the union of events like HH ... HM ... M, where this is a sequence containing x hits and $n-x$ misses in a definite order. Each of these events will have probability $p^x(1-p)^{n-x}$, since the probability of a hit is p and the probability of a miss is $1-p$. There will be altogether $\binom{n}{x}$ of these events, this being the number of ways in which we can choose the x shots that are going to be hits.

Hence $P(x) = \binom{n}{x} p^x(1-p)^{n-x}$ for $x = 0, 1, \ldots, n$.

This is often written instead as

$$P(x) = \binom{n}{x} p^x q^{n-x} \text{ where } q = 1-p.$$

(We can check that this formula gives the results we had earlier for

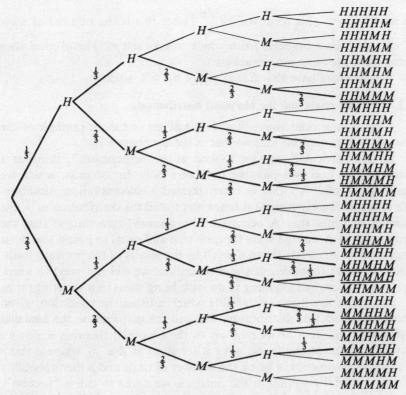

Figure 8.3
Probability tree for George's distribution showing $P(2) = 10(\frac{1}{3})^2 \cdot (\frac{2}{3})^3$

$n = 5$, remembering that $\binom{n}{0} = \binom{n}{n} = 1$.)

We can again verify directly that $\Sigma P(x) = 1$.

For $\Sigma P(x) = \sum_{x=0}^{n} \binom{n}{x} p^x q^{n-x}$

$$= q^n + \binom{n}{1} q^{n-1}p + \binom{n}{2} q^{n-2}p^2 + \ldots + p^n.$$

However, by the binomial theorem,

$$(q+p)^n = q^n + \binom{n}{1} q^{n-1}p + \binom{n}{2} q^{n-2}p^2 + \ldots + p^n.$$

(We can in any case see that this last result is true by writing $(q+p)^n$ as $(q+p)(q+p)(q+p)\ldots(q+p)$. Every term in the expansion must consist of the product of n factors, each of which is either a p or a q, and could therefore be written as $q^{n-x}p^x$ for some x, $0 \leqslant x \leqslant n$. The number of terms of exactly this form will be $\binom{n}{x}$ since this is the number of ways of choosing the x brackets from which to take the ps. The qs must then come from the remaining brackets.)

We therefore have that $\Sigma P(x) = (q+p)^n = \cdot 1$, since $q = 1-p$.

8.2. Repeated trials and the binomial distribution

What is so special about this? Is it just yet one more example of the sort of random variable that we met in the last chapter?

In this particular case we looked at an "experiment", firing at a target, which had two possible outcomes only, hit or miss, which we shall call A and A′; this is often termed a success/failure situation. We repeated the experiment n times and found the distribution of x, the number of times that A occurred. We assumed two things: that the probability of A was the same for each trial and equal to p, and that what happened on any trial was unaffected by the results of the previous trials. Then, in order to work out the distribution, we did not need to know that our problem had anything to do with firing shots in a shooting range.

So we can say that any variable which satisfies the conditions given above will have this distribution. We call the distribution the **binomial distribution** (because of its relation to the binomial theorem which we have already mentioned) and write it for short as $B(n, p)$, where n and p are its two *parameters*, n being the number of trials and p the probability of any one trial resulting in the outcome we agree to call a "success". Notice that the whole distribution is specified, provided we know the values of n and p, as the parameters give us all the possible information about the distribution. For instance the distributions $B(5, \frac{1}{2})$ and $B(5, \frac{1}{3})$ are given in the first two diagrams in this chapter.

You can probably think of other examples in which the binomial distribution will hold. One obvious one is counting the number of heads obtained in n tosses of a coin (or, which is exactly the same thing, the simultaneous tossing of n exactly similar coins). We have already looked at this problem, worked out in a slightly different way, in Chapter 2, when we considered George's problem of whom to take to the cinema. There, since he tossed 3 fair coins, his variable (the number of heads) had a $B(3, \frac{1}{2})$ distribution.

For a more practical example we could consider a factory making gramophone records of which 5% have defective mouldings. The factory inspector takes a daily sample of 40 records, chosen at random from the day's production, and counts the number of defective records, x say, in each sample.

Let us suppose that the day's production of 10 000 records contained 500 defectives. If we assume that the sample is taken *with replacement*, each record in the sample will have the same probability 0·05 of being defective and x will have the $B(40, 0·05)$ distribution. If, as is more likely, the sampling is *without replacement*, the probability of being defective will change slightly for each successive record in the sample and x will strictly have a different distribution, the *hypergeometric distribution* (Exercise 8A, question 21). However, since the sample is very small compared with the day's production, it can be shown that x is still approximately $B(40, 0·05)$. (In general we can assume that sampling without replacement from a very large population will give approximately the binomial distribution.)

What is the probability that the inspector finds 3 or more defective records in a day's sample?

$P(x \geqslant 3) = 1 - P(x \leqslant 2) = 1 - P(0) - P(1) - P(2)$
$= 1 - (0·95)^{40} - \binom{40}{1} \times (0·95)^{39} \times (0·05) - \binom{40}{2} \times (0·95)^{38} \times (0·05)^2 \simeq 0·32$

8.3. The mean and variance of $B(n, p)$

In the shooting example George was also interested in their average number of hits per round. Jim calculates that, as he hits the ping-pong ball once out of every 2 turns, he could reasonably expect to hit it on average 2·5 times in a round of 5 turns. Similarly George, whose probability is $\frac{1}{3}$ would expect to hit the ball once in every 3 turns, giving him an average of $\frac{5}{3}$ hits in a round of 5 turns. You can verify these results from the probability functions.

Similarly, with a probability of 0·05 of getting a defective record, in a sample of 40 records most people would expect to get an average of 2 defectives. If we generalized this argument, we would probably guess that the mean of the binomial distribution $B(n, p)$ was np. If we look more closely at the example of the records, perhaps we can see why our guess is correct.

$E(x) = \Sigma x P(x)$

So here $E(x) = 0 . P(0) + 1 . P(1) + 2 . P(2) + \ldots + 39 . P(39) + 40 . P(40)$

$= 0 + 1\binom{40}{1}(0·95)^{39}(0·05) + 2\binom{40}{2}(0·95)^{38}(0·05)^2 + 3\binom{40}{3}(0·95)^{37}(0·05)^3$

$\quad + \ldots + 39\binom{40}{39}(0·95)^1(0·05)^{39} + 40(0·05)^{40}$

$= 40(0·95)^{39}(0·05) + \frac{40.39}{1}(0·95)^{38}(0·05)^2 + \frac{40.39.38}{1.2}(0·95)^{37}(0·05)^3 +$

$\quad \ldots + 39 . 40(0·95)^1(0·05)^{39} + 40(0·05)^{40}$

$= 40(0·05)[(0·95)^{39} + 39(0·95)^{38}(0·05) + \frac{39.38}{1.2}(0·95)^{37}(0·05)^2 + \ldots$

$\quad + 39(0·95)^1(0·05)^{38} + (0·05)^{39}]$

$= 40(0 \cdot 05) \ [0 \cdot 95 + 0 \cdot 05]^{39}$ using the binomial theorem
$= 40(0 \cdot 05) = 2.$

We can see that we could prove this just as easily in the general case with the result $E(x) = np(q+p)^{n-1} = np$, simply by replacing 40 by n, 0·05 by p and 0·95 by q.

Thus, as we guessed, the mean of the binomial distribution $B(n,p)$ is np.

What about the variance? If we return to Jim and George again, we can say that

$\sigma^2{}_{\text{JIM}} = \Sigma x^2 P(x) - \mu^2 = \tfrac{1}{32}(0 + 1 \times 5 + 4 \times 10 + 9 \times 10 + 16 \times 5 + 25 \times 1) - \tfrac{25}{4}$
$= \tfrac{5}{4}$

We can work out $\sigma^2{}_{\text{GEORGE}}$ similarly; in fact it equals $\tfrac{10}{9}$.

Have these any connection with $E(x)$ in each case? For Jim we have $E(x) = \tfrac{5}{2}$ and $\sigma^2 = \tfrac{5}{4}$, therefore $\sigma^2{}_{\text{JIM}} = \mu_{\text{JIM}} \times \tfrac{1}{2}$; similarly $\sigma^2{}_{\text{GEORGE}} = \mu_{\text{GEORGE}} \times \tfrac{2}{3}$. But $\tfrac{1}{2} = 1 - p_{\text{JIM}}$, $\tfrac{2}{3} = 1 - p_{\text{GEORGE}}$. So in both cases we have $\sigma^2 = np(1-p) = npq$.

It is possible to derive this result generally in much the same way as we calculated the mean, though the proof is rather complicated. We shall leave the direct proof as an exercise and prove the result using probability generating functions in the next section.

8.4. The probability generating function

In the last chapter we defined the probability generating function (p.g.f.) of a random variable to be $G(t) = E(t^x) = \Sigma t^x p_x$ for a variable taking only integer values.

Let us find the p.g.f. for the binomial distribution $B(n,p)$.

$G(t) = \overset{n}{\underset{x=0}{\Sigma}} t^x \binom{n}{x} p^x q^{n-x} = \overset{n}{\underset{x=0}{\Sigma}} \binom{n}{x} (pt)^x q^{n-x} = (q+pt)^n.$ So in this case the p.g.f. has a very simple form. Putting $t = 1$ gives us $G(1) = (q+p)^n = 1$ which confirms our earlier calculations.

$G'(t) = np(q+pt)^{n-1} \quad \therefore \mu = G'(1) = np(q+p)^{n-1} = np.$
$G''(t) = n(n-1)p^2(q+pt)^{n-2}$
$\therefore \sigma^2 = G''(1) + G'(1) - (G'(1))^2 = n(n-1)p^2 + np - (np)^2$
$= np - np^2 = np(1-p) = npq$

So $\sigma^2 = npq$, which is what we wanted to prove earlier.

This gives the variance of the number of defective records per sample as $40 \times (0 \cdot 05) \times (0 \cdot 95) = 1 \cdot 90.$

EXERCISE 8A

1. A fair coin is spun five times. Find the probability that:
   (i) there will be exactly two heads,

(ii) there will be at least two heads,

(iii) no two consecutive spins will have the same result. (J.M.B.)

2. Find the probability that of the first six people met in the street on a given day, at least four will have birthdays on a Sunday. Give your answer as a fraction. ($7^6 = 117,649$.) (S.M.P.)

3. The chance that the server at tennis wins a particular point is $\frac{2}{3}$. Find the chances that, of four consecutive points, the server (a) wins all four, (b) wins exactly three, (c) loses at least three. (Cam.)

4. 75% of the very large number of candidates in a public examination pass. If the chief examiner selects at random ten marked scripts, show that the probabilities that he finds among the ten (a) only two passes, and (b) no more than two passes, are just less than and just greater than 4×10^{-4} respectively. (S.M.P.)

5. Draw the frequency diagram of the distribution $B(n, p)$ for the cases: (i) $p = 0.3, n = 8$, (ii) $p = 0.5, n = 10$, (iii) $p = 0.1, n = 5$, (iv) $p = 0.1, n = 10$.

6. When Leeds United are at home to Liverpool, it is estimated that for every Liverpool supporter in the crowd there are six Leeds supporters. Assuming that 8 people are charged for bad behaviour during the match, what is the probability that more than 3 of them are Liverpool supporters? Do you think your model is a good one in this situation?

7. If it can be assumed that the numbers of boys and girls that are born are approximately equal, find the proportion of families of six children that:

(a) contain as many boys as girls,

(b) are all of the same sex.

8. In a row of 7 telephone kiosks at a station there are on average 2 free at any time during the rush hour. What is the probability that:

(a) there will be just one free at a given time,

(b) you will have to wait for a free kiosk,

(c) there will be more than one free? Is your model a good one?

9. After fireworks have been stored in a certain climate it is found that an average of one fifth of them will not light. A boy buys three. What is the probability that exactly two of them will light? A man wishes to be 95% sure (i.e. have a probability ≥ 0.95) of having at least two which will light. Investigate whether four will be enough for him to buy. (S.M.P.)

10. A man tosses a penny eight times. For each head he moves 1 m northwards and for each tail 1 m southwards. Is the chance that he will end up at his starting-point:

(a) less than $\frac{1}{8}$,

(b) between $\frac{1}{8}$ and $\frac{1}{4}$,

(c) between $\frac{1}{4}$ and $\frac{1}{2}$,

(d) greater than $\frac{1}{2}$?

Give reasons for your answer. Estimate the probability that he will end up more than 3 m away from his starting-point. (Cam.)

11. For the distribution $B(n, p)$, find an expression for $P(x)$ in terms of $P(x-1)$. A computer installation has 5 terminals. The probability that any one particular terminal will not require attention during a shift is 0.85. Find the probability that the engineer will not be called out during the shift. Hence, using the formula you derived in the first part of the question, find the probabilities that the engineer will have to attend to 1, 2, 3, 4, or all of the terminals.

12. Part of an examination consists of 14 multiple-choice questions each with four alternatives and each allotted 1 mark. What is (a) the mean mark, (b) the most probable mark, of a candidate who just guesses? If it is decided to subtract one mark for every incorrect answer, how many marks should be given for each correct answer to make the mean mark for these candidates zero?

13. Using the formula you derived in question 11, find for what range of x, $P(x)$ is greater than $P(x-1)$ for the distribution $B(n, p)$. Can you say anything about the mode of the distribution (the most probable value):
   (a) when $(n+1)p$ is not a whole number,
   (b) when $(n+1)p$ is a whole number?

14. By rewriting the proof in § 8.3 in terms of n, p and q, prove that the mean of the distribution $B(n, p)$ is np. Use a similar method to show that if x has this distribution $E(x(x-1)) = n(n-1)p^2$ and hence verify the formula for Var(x).

15. On my journey to work I have to pass through 12 sets of traffic lights which operate independently; the chance that any set is at green when I reach it is $\frac{1}{2}$. If I am stopped at fewer than 3 sets, I have time for a cup of coffee before work, but if I am stopped at more than 8 sets I am late. If I am late more than twice in a five-day week, I have to see the Office Manager. What is the probability that:
   (a) I have time for coffee on two given consecutive mornings,
   (b) I have to see the Office Manager in the week after my holiday?   (Cam.)

16. Two players stake £1 each in a game in which eight counters are tossed on the ground. The counters are painted black on one side and white on the other. If the number of blacks showing is odd, the thrower wins the other stake. If all counters are of the same colour, he wins a double stake; and otherwise he loses his stake. Calculate the expected gain of the thrower, and state what assumptions you make.   (M.E.I.)

17. On average 5% of the eggs supplied to a supermarket are cracked. If you buy ten boxes each containing six eggs, what is the probability that none of your boxes contain two or more cracked eggs? What is the mean number of boxes that contain two or more cracked eggs?

18. In a certain competition for teams of six the qualifying event consists of each member of the team being allowed up to three attempts at a particular trial, the team as a whole qualifying if at least four members out of the six succeed in the trial. Assuming that the chance of failure in any one attempt by any member of the team is q, evaluate the chance of the team qualifying.   (Cam.)

19. In the next general election suppose that 36% of the electors intend to vote Liberal. Find, in terms of n, the mean and the standard deviation of the percentage of those who intend to vote Liberal in samples of size n.   (S.M.P.)

20. Playing a certain "one-armed bandit", which is advertised to "Increase your money tenfold" costs $2\frac{1}{2}$p a turn; the player is returned 25p if more than 8 balls out of a total of 10 drop in a specified slot. The chance of any one ball dropping is p. Determine the chance of winning in a given turn, and for $p = 0.65$ calculate the mean profit made by the machine on 500 turns. Evaluate the proportion of losing turns in which the player comes within one or two balls of winning $(p = 0.65)$.   (Cam.)

21. A bag contains N balls of which R are red and the remainder white. A sample of n balls is drawn from the bag at random and without replacement. It may be assumed that $n \leqslant R$ and $n \leqslant N - R$. Show that the probability that the sample contains r red balls is given by

$$P(r) = \frac{\binom{R}{r}\binom{N-R}{N-r}}{\binom{N}{n}} \qquad 0 \leqslant r \leqslant n.$$

Show that $rP(r)$ may be written as

$$\frac{nR}{N} \cdot \frac{\binom{R-1}{r-1}\binom{N-1-(R-1)}{N-1-(r-1)}}{\binom{N-1}{n-1}}$$

and hence find the mean number of red balls in the sample. Find a similar expression for $r(r-1)\,P(r)$ and show that the variance of the number of red balls may be written as

$$\frac{n(N-n)}{(N-1)}\,pq \text{ where } p = \frac{R}{N} \text{ and } q = 1 - p.$$

This distribution is known as the *hypergeometric distribution*. When N is large it may be approximated by the binomial distribution $B(n,p)$. Show that for N large the mean and variance obtained tend to np and npq respectively. Describe how this result could be used to estimate the size of an animal population using a 'ringing' or 'marking' system.

8.5. A limit of the binomial distribution

Suppose we consider a second factory, but this time one involved in weaving cotton material. For a certain loom the finished material is found to have on average one fault in every 10-metre length of cloth. If 10-metre lengths are taken from this loom, what is the distribution of the number of faults in them?

This time we are dealing with a rather different problem in that there is no clear limit to the number of faults which might occur; it might be any non-negative integer in a very wide range.

To simplify the problem let us suppose that we have divided the 10-metre length into 20 small pieces, each 0·5 m long. We will make two assumptions about these small pieces; first, that no more than one fault can occur in each small piece, and second, that the number of flaws in any one piece is not affected by the number of flaws in any other small piece. Now we are back in a familiar situation, for, if x is the total number of flaws in the 10-metre length and p is the probability of a flaw in any one of the small pieces, then x will have the $B(20, p)$ distribution. We can go further and find p. We know that the average number of flaws in a 10-metre length is 1, and therefore we must have $20p = 1$ and therefore $p = 0·05$. So under these assumptions x has the distribution $B(20, 0·05)$.

Are the assumptions realistic? The second assumption of independence is a necessary one if we are going to use any model based on the binomial distribution; if we suspect this assumption is not true, the model must be changed. The first assumption, however, can be refined. If we think it possible to get more than one fault in a piece of cloth 0·5 m long, we can divide our 10-metre length into 50 small pieces, each 0·2 m long. In that case x would have the distribution $B(50, 0·02)$. (We must still ensure that the mean number of faults in the whole piece is 1.)

Even these pieces might not be small enough; we might instead take 100 small pieces 0·1 m long, giving x the distribution $B(100, 0·01)$. The larger the number of small pieces we take the more likely it is that our first assumption will be justified. What we really need to do is to let the number of pieces into which we divide our length get larger and larger and see if the distribution tends to a limit.

Let us start by summarizing the distributions we have suggested.

p	n	np	npq
0·05	20	1	0·95
0·02	50	1	0·98
0·01	100	1	0·99

We arranged that all three distributions should have a mean of 1. Notice that the variances are approximately equal as well and look as though they are approaching the value 1 as n increases. We might suspect that since the distributions are all binomial, have the same mean, and approximately the same variance, then they might have approximately the same distribution as well. If we work out $P(2)$ in all three cases we find:

$$B(20, 0·05): \ P(2) = \binom{20}{2}(0·05)^2 (0·95)^{18} = 0·1887$$
$$B(50, 0·02): \ P(2) = \binom{50}{2}(0·02)^2 (0·98)^{48} = 0·1858$$
$$B(100, 0·01): \ P(2) = \binom{100}{2}(0·01)^2 (0·99)^{98} = 0·1849$$

So this at least suggests that if we go on increasing the number of divisions (and decreasing the probability p in order to keep the mean constant) the distributions we obtain might tend to a limit, which is the distribution we want. Let us see if we can prove this mathematically and find the form of the distribution.

What we want to do is take the binomial distribution given by $P(x) = \binom{n}{x}p^x q^{n-x}$ and let n increase and p decrease in such a way that we always have $np = 1$. Since $np = 1$ we have $p = 1/n$ and we can rewrite $P(x)$ as

$$P(x) = \binom{n}{x}\frac{1}{n^x}\left(1 - \frac{1}{n}\right)^{n-x} = \frac{n(n-1)\dots(n-x+1)}{x!\,n^x}\frac{(1-\frac{1}{n})^n}{(1-\frac{1}{n})^x}$$

$$= \left\{ \frac{1(1-\frac{1}{n})\dots(1-\frac{x-1}{n})}{x!} \right\} \frac{(1-\frac{1}{n})^n}{(1-\frac{1}{n})^x}$$

As n gets large $(1-1/n)$ and $(1-2/n)$ ... and $(1-(x-1)/n)$ all tend to 1, so that the expression in curly brackets will get closer and closer to $1/x!$. Since x is a fixed number, $(1-1/n)^x$ will also tend to 1 as n gets large (see the table below for an illustration in the case $x=2$). What about $(1-1/n)^n$ though? As we have just seen $(1-1/n)^x$ gets close to 1, but as our exponent n is increasing at the same time, this is no help to us. The second line of the table below gives values of $(1-1/n)^n$ for various values of n.

n	2	3	4	5	10	20	50	100	500	1000	5000
$(1-\frac{1}{n})^2$	0·250	0·444	0·563	0·640	0·810	0·903	0·960	0·980	0·996	0·998	1·000
$(1-\frac{1}{n})^n$	0·250	0·296	0·316	0·328	0·349	0·359	0·364	0·366	0·368	0·368	0·368

The successive terms are increasing and seem to be getting closer to a limiting value c as n increases. From the table we see that c is approximately equal to 0·368.

So if we substitute all the limiting values in the expression for $P(x)$, we have $P(x)\to c/x!$ as n gets larger. What values can x take? Originally x was restricted to taking integer values between 0 and n, but as we have now let n get large, it seems clear that x must take all integer values $\geqslant 0$. Notice that this is just what we said earlier about the possible values of the number of flaws in our length of material.

So, we have derived a new distribution $P(x) = c/x!$ where $x = 0, 1, 2, 3, \dots$, which will give us the distribution of the number of flaws in a length of material. Of course we should check that $P(x)$ satisfies the condition $\Sigma P(x) = 1$. $\Sigma P(x) = c\Sigma(1/x!)$.

But $\Sigma \dfrac{1}{x!} = 1 + \dfrac{1}{1!} + \dfrac{1}{2!} + \dfrac{1}{3!} + \dots = 1+1+0·5+0·16+\dots = 2·718\dots$.

This is a number commonly met in mathematics and is called e. It is similar to π in that neither can be expressed as a simple fraction. We notice that $c = \lim(1-1/n)^n \simeq 0·368$ and $1/e \simeq 1/2·718 \simeq 0·368$, so it looks as though $c = 1/e$. In fact it can be proved, though we shall not do it here, that $\lim(1-1/n)^n = e^{-1}$ so that we have $\Sigma P(x) = \Sigma(e^{-1}/x!) = e^{-1}e$, which satisfies the condition $\Sigma P(x) = 1$.

So we can say that the limiting form of our probability distribution for the number of flaws in our length of cloth is $P(x) = e^{-1}/x!$, $x = 0, 1, 2, \dots$ (If you substitute $x = 2$ in this expression you obtain $P(2) = e^{-1}/2 = 0·184$, which agrees very well with our earlier approximate results.)

It should be clear that we can employ this result and say that if we have a binomial distribution and the number of trials n is fairly large, and if the mean is 1, then $P(x)$ is approximately equal to $e^{-1}/x!$, which is in fact independent of the parameters n and p. We will discuss situations when we can use this result in the next section.

What would have happened if the average had been 2 or 3 instead of 1? In general let us suppose that the average was μ. We can find the limit of $P(x)$ in just the same way as before, though now we shall have $p = \mu/n$ to keep the average μ. If we write out the expression for $P(x)$ as we did before, we will see that it is almost identical except for an extra term μ^x in the numerator and now we shall have to write $c_\mu = \lim(1-\mu/n)^n$ to replace c. So we can write $P(x)$ as $P(x) = c_\mu \mu^x/x!$, $x = 0, 1, 2 \ldots$

Now we need to find $\lim (1-\mu/n)^n$. The table below shows the situation for $\mu = 2$ and 3.

n	2	3	4	5	10	20	50	100	500	1000	5000
$(1-\frac{2}{n})^n$	0·000	0·037	0·063	0·078	0·107	0·122	0·130	0·133	0·135	0·135	0·135
$(1-\frac{3}{n})^n$	0·250	0·000	0·004	0·010	0·028	0·039	0·045	0·048	0·049	0·050	0·050

We notice that $c_2 \simeq 0·135 \simeq c^2 = e^{-2}$ so that $\lim \left(1 - \dfrac{2}{n}\right)^n = e^{-2}$,

$c_3 \simeq 0·050 \simeq c^3 = e^{-3}$ so that $\lim \left(1 - \dfrac{3}{n}\right)^n = e^{-3}$.

So we might postulate that, in general $c_\mu = \lim(1-\mu/n)^n = e^{-\mu}$. We can confirm this by using the fact that we must have $\Sigma P(x) = 1$.

$$\Sigma P(x) = \Sigma \frac{c_\mu \, \mu^x}{x!} = c_\mu \Sigma \frac{\mu^x}{x!}$$

The result $\Sigma(\mu^x/x!) = 1 + \mu + \mu^2/2! + \ldots = e^\mu$ can easily be proved by other means and this verifies the result $c_\mu = e^{-\mu}$.

The distribution becomes $P(x) = e^{-\mu} \mu^x/x!$. This would give us the

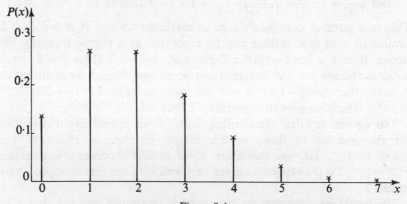

Figure 8.4
Poisson distribution ($\mu = 2$)

probability distribution of the number of flaws in a 10-metre length of cloth (or indeed in any other length) if the mean number of flaws in that length was μ. The shape of the distribution is shown for $\mu = 2$ in Fig. 8.4.

8.6. The Poisson distribution

In the last section we found what happened to the binomial distribution when we fixed its mean np and let n get large. We found it tended to another distribution $P(x) = e^{-\mu} \mu^x/x!$ which we call the **Poisson distribution,** after its originator and which we write as $\mathscr{P}(\mu)$. We will use $\mathscr{P}(\mu)$ as an approximation to the binomial distribution when we have $np = \mu$ and n large and p small. Its terms can be easily calculated using the table of e^{-x} on page 247. It is normally safe to use this approximation for $n > 40$ provided p is small ($< 0\cdot1$ say) but the larger n, the more accurate the approximation.

We will thus use the distribution to describe the probability of the occurrence of rare events as in the next two examples. Think of all the passengers on London Transport buses on any day; clearly a large number, n say. Suppose that the probability that a passenger leaves something on the bus is p (assuming it is the same for all passengers which is reasonably close to the truth), then p will be very small, but the average number of articles left on the buses in a day is likely to be considerable. If this average is μ then we could approximate to the distribution of the number of articles left on any one day by $\mathscr{P}(\mu)$. Or, suppose a tape-punch operator is punching data tapes containing 500 symbols. On average she makes one mistake every 100 symbols. The distribution of the number of mistakes on a tape would be given approximately by $\mathscr{P}(5)$, provided we assume that she does not have good and bad patches of punching. Thus, the probability that she makes no mistakes on a tape would be given by $P(0) = e^{-5}5^0/0! = e^{-5} \simeq 0\cdot007$, and the probability that she makes five or more mistakes is given by

$$P(5) + P(6) + P(7) + \ldots = 1 - P(0) - P(1) - P(2) - P(3) - P(4)$$
$$= 1 - e^{-5} - e^{-5}.5 - e^{-5}.5^2/2! - e^{-5}.5^3/3! - e^{-5}.5^4/4! \simeq 0\cdot560.$$

For another example we might consider cars passing along a road at an average of 240 vehicles per hour. What is the probability distribution of the number passing in a given minute? On average $\frac{240}{60} = 4$ will pass in one minute. So we could make the same assumption as we did for the material, that the number passing in any small interval of time is independent of the number passing in any other interval of time, and deduce that the distribution is $\mathscr{P}(4)$ by a similar argument. The assumption may easily not be justified here, particularly considering such things as the queues which follow a slow-moving vehicle, but in practice the Poisson distribution provides a good model of the situation.

From the way we derived the Poisson distribution it is clear that its mean must be μ. We can check this as follows:

$$E(x) = \Sigma x P(x)$$
$$= \sum_{x=1}^{\infty} x e^{-\mu} \frac{\mu^x}{x!} = \mu e^{-\mu} \sum_{x=1}^{\infty} \frac{\mu^{x-1}}{(x-1)!}$$
$$= \mu e^{-\mu}\left(1 + \mu + \frac{\mu^2}{2!} + \ldots\right) = \mu.$$

What about the variance? We derived the Poisson distribution from the binomial distribution for which $\sigma^2 = npq$, by setting $np = \mu$ and letting $n \to \infty$, $p \to 0$ and therefore $q \to 1$. This suggests that for the Poisson distribution

$$\sigma^2 = \lim (np)q = \mu.$$

We could check this result either from the probability distribution using the formula $E(x^2) - \mu^2$, or from the probability generating function but we will leave these as exercises.

EXERCISE 8B

1. Cars arrive at a motorway garage at an average rate of one every minute. A crisis occurs if more than four cars arrive in any one minute. How many crises are there on average in a twelve-hour day? Do you think that the distribution you have used is a good model in this case?

2. The *Wickington Chronicle* has an average of 3 misprints per page. What is the probability that:
   (a) the front page is free from misprints,
   (b) there are exactly 6 misprints on the back page,
   (c) there are more than 3 misprints on the sporting page?

3. An insurance company finds that 0·005 per cent of the population dies from a certain kind of accident each year. Ten thousand people are insured with the company against this risk. Calculate (to 2 significant figures) the probability that the company will receive at least 4 claims for this accident in any given year. (J.M.B.)

4. An airline finds that, on average, 4 per cent of the persons who reserve seats for a certain flight do not, in fact, turn up for the flight. Consequently the airline decides to allow 75 persons to reserve seats on a plane which can only accommodate 73 passengers. What is the probability that there will be a seat available for every person who turns up for the flight? (J.M.B.)

5. The road accidents in a certain area occur at an average rate of one per two days. Calculate the probability of 0, 1, 2, ..., 6 accidents per week in the district. What is the most likely number of accidents per week? How many days in the week are expected to be free of accidents? (M.E.I.)

6. A shop sells radios of a particular make at an average rate of four per week. Assuming that the number sold in a week is a Poisson variate, find what number should be in stock at the beginning of a week so as to have a 95% assurance of being able to meet all demands during the week. (M.E.I.)

7. Of 150 football matches played last Saturday there were 12 in which there was no score. Assuming a Poisson distribution, what do you think the mean number of goals per match was? Find the probabilities that:
   (a) less than 2 goals were scored in a match,
   (b) more than 2 but less than 5 goals were scored.

8. Find the probability generating function of the Poisson distribution $\mathscr{P}(\mu)$ and verify the results for the mean and variance obtained in the chapter.

9. By investigating the relationship between $P(x)$ and $P(x-1)$, find the mode(s) of $\mathscr{P}(\mu)$ and the condition for it to be bimodal.

10. For the distribution $\mathscr{P}(\mu)$ show that $\Sigma x(x-1) P(x) = \mu^2$ and hence find the variance.

11. Vehicles pass a point on a busy road at an average rate of 300 per hour. Find the probability that none pass in a given minute. What is the expected number passing in two minutes? Find the probability that this expected number actually pass in a given two-minute period.

12. The number of accidents per day was recorded in a certain district for a period of 1500 days and the following results were obtained:

Number of accidents per day	0	1	2	3	4	5
Frequency	342	483	388	176	111	0

What theoretical distribution may be used as a suitable model for these data? Calculate the expected frequencies based on the theoretical probability distribution with mean equal to the average of the observed values. (M.E.I.)

13. In question 7, assuming that on average the home team scores twice as many goals as the away team, and that the scores of the two teams are independent, what do you think the distribution of the number of home goals might be? Can you prove your result? Show that the probability of the game being drawn is
$$\sum_{x=0}^{\infty} \frac{2^{x+1}}{25(x!)^2} (\tfrac{2}{3} \ln 5 - \tfrac{1}{3} \ln 2)^{2x}.$$

14. The probability of more than one fatal accident in any month in Munchester is 0·1. Assuming that the number of fatal accidents has a Poisson distribution, draw a graph of the function $y = (x+1)e^{-x}$ and estimate the mean number of accidents per month. Show that, if you are told the probability of exactly one fatal accident instead, the mean is not uniquely determined.

15. Jim goes fishing every Saturday but on a proportion p of these days he does not catch anything. He makes it a rule to keep the first and then every alternate fish that he catches. Find:
   (a) the mean number of fish he catches,
   (b) the probability that he takes home x fish,
   (c) the probability that he takes home the last fish he catches.
   Prove that the mean number of fish he takes home is $\tfrac{1}{4}[2 \ln(1/p) + 1 - p^2]$.

Continuous Variables

9.1. A new look at the uniform distribution

If we ask a group of people to write down a whole number between 0 and 9 inclusive, we may well find that the choices appear not to be very random; there are one or two numbers which are often chosen much more frequently than the others.

If, however, we are investigating a situation in which we feel we are justified in assuming the numbers to be randomly chosen (for instance by computer or by pulling them out of a hat) we can easily write down the probability distribution, which would predict the actual frequency distribution we might expect to obtain from a large number of trials.

There are clearly ten possible values, each of which we are assuming to be equally likely, and hence $P(x) = \frac{1}{10}$ for $x = 0, 1, 2, \ldots 9$ and $P(x) = 0$ for x outside the range 0–9. The distribution, very simply, would look like Fig. 9.1.

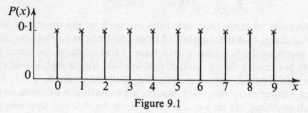

Figure 9.1

The distribution is uniform and its cumulative distribution function would look like Fig. 9.2.

Suppose we now alter the problem somewhat and allow the number chosen x to be any real number between 0 and 10, including 0 but not 10. In practice this would be difficult as most random-number devices are unable to choose a number like π which cannot be expressed as a finite decimal, but we will assume that the experiment could be carried out. We would now find it impossible to draw a probability chart of the situation,

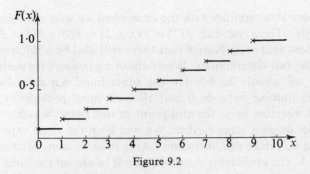

Figure 9.2

as we would have to draw a column for each real number in the interval. Instead let us group the numbers into the ranges 0–, 1–, 2–, . . . , 9– and consider the probability of x lying in each of these intervals. As we are assuming x to be chosen at random from the interval 0–10, the probability of x lying in any one of these small intervals will still be $\frac{1}{10}$. So instead of drawing a probability chart we can represent our new experiment in a histogram as shown in Fig. 9.3.

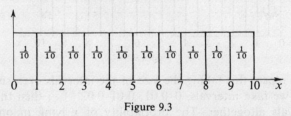

Figure 9.3

We can find the cumulative distribution function for our new experiment too. We have, for example, $F(2) = P(x \leqslant 2) = \frac{2}{10} = 0.2$ and for any integer r, $0 \leqslant r \leqslant 10$, $F(r) = r/10$. A little thought shows us that this expression is not only going to hold for integer r but for all real r. For instance $F(3.2) = 0.32$. A graph of the cumulative distribution function looks like Fig. 9.4.

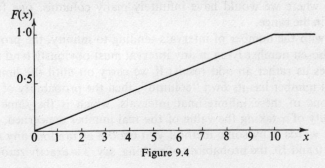

Figure 9.4

Compare this carefully with the case when we were choosing integer values only. There we had $F(2) = P(x \leqslant 2) = P(0) + P(1) + P(2) = \frac{3}{10}$; whereas now $F(2) = \frac{2}{10}$. Notice that there will also be a difference in the mean of the two distributions. When choosing integers between 0 and 9, the mean will clearly be 4·5. On the other hand our new distribution allows any number between 0 and 10 with equal probability, and the mean will therefore lie at the mid-point of this range, which is 5.

Now let us go a stage further. We will keep the same experiment as before, but now draw our histogram taking twice as many intervals, 0, 0·5, 1·0, . . ., 9·5. The probability distribution will be almost the same as before except that each small interval will have a probability of $\frac{1}{20}$ associated with it. In the histogram the area of each column must represent the probability of the corresponding interval; so since each interval has length 0·5, the height of each column will still be $\frac{1}{10}$. The cumulative distribution functions will be exactly the same in the two cases.

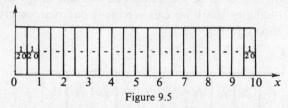

Figure 9.5

We can repeat this taking the size of the intervals even smaller. For instance if we take intervals, 0–0·01, 0·01–0·02, . . ., then there will be 1000 intervals altogether. The probability of x lying in one of these intervals is therefore $\frac{1}{1000} = 0.001$, but when we come to draw the histogram each column will have height $\frac{1}{10}$ as before. Notice that in all three cases the total area of the histogram is 1.

Suppose we continue the process with progressively narrower intervals. Each time the probability of x lying within any interval must get smaller and smaller, and we would gradually approach the limiting position where we would have infinitely many columns, one for each number in the range.

But, with the number of intervals tending to infinity, the probability of our chosen number lying in any interval must obviously tend to zero. This gives us rather an odd result; if we carry on until the limit when each real number has its own "column", then the probability of x lying in any one of these infinitesimal intervals, which is the same as the probability of x taking the value of the real number concerned, is zero. In other words if by some method you choose at random any number between 0 and 10, the probability of picking, say, 3 is exactly zero.

Although this seems odd it must be so. For suppose the probability of picking 3 was p, then the probability of picking any other number must also be p, so that if we added up the probabilities over infinitely many numbers, the total certainly could not equal 1, which is the condition for a probability distribution.

This means that we will have to modify our ideas about zero probability. We started off with the idea that an event had probability zero only if it was impossible. We must now allow this also to include some events which *are* possible, provided they occur as the result of experiments with an infinite number of outcomes. Notice however that not all experiments with infinitely many outcomes need have this property; in the example of waiting for a son which we met in Chapter 7, the number of children could be any whole number but each number had a non-zero probability.

We notice that, however small we make the intervals, the height of the column will still be $\frac{1}{10}$ in order to keep the total area (which represents the sum of the probabilities) equal to 1. This height therefore certainly does

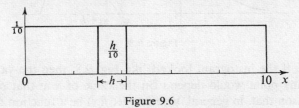

Figure 9.6

not represent the probability of any column; but it is connected to it in that if the interval is of width h, then the probability of the number lying within that interval is equal to the area of the column, which is $h/10$.

So in the continuous case, when we consider every possible value of x over a range, we are bound to meet the fact that the probability of x taking any particular value is zero, but the probability of it lying within any particular interval within the range is a number greater than zero.

9.2. The probability density function

From the results of the last section, we also have a new variable, the height of the histogram, which we shall call the **probability density function** (p.d.f.) and denote by $f(x)$.

The p.d.f. clearly has links with the *frequency density* which is represented by the vertical scale in an experimentally obtained histogram. One difference is that in the case of a histogram the total area represents not 1 but the total frequency n, but if the vertical scale is reduced by a factor n to give the *relative frequency density* the analogy is exact.

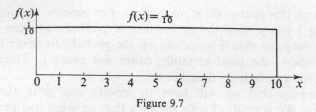

Figure 9.7

For our uniform distribution we have
$$f(x) = \tfrac{1}{10}, 0 \leqslant x \leqslant 10$$
$$= 0, x < 0, x > 10$$
(Often the second part of this definition is omitted and it is understood that $f(x) = 0$ for all undefined values of x.)

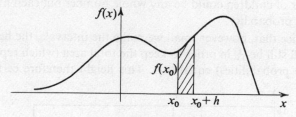

Figure 9.8

However if the histogram looked like Fig. 9.8, then the value of the height at any point would depend on the value of x at that point. It is for this reason that, in general, we must let $f(x)$ be a function depending on x.

We have seen that, for the uniform distribution,
$$P(x \text{ lies in an interval of length } h) = h/10 = f(x).h$$
This is equal to the area under the curve between the end-points of the interval.

Now we wish to see how we can extend this result to probability density functions like that drawn above. For any curve representing a probability density function, the probability that x lies in any small interval will be given by the area under the curve between the end points of the interval.

So we see from the diagram, if we choose a *small* interval (x_0, x_0+h), the probability that x lies in this interval will be approximately $f(x_0).h$. Or, if we use the traditional symbol δx instead of h, we can write $P(x_0 \leqslant x \leqslant x_0+\delta x) \simeq f(x_0).\delta x$ provided δx is sufficiently small. The two sides will become more nearly equal as $\delta x \to 0$.

You will realize that we are now into the realms of calculus. $P(a \leqslant x \leqslant b)$ is the area under the curve $y = f(x)$ between the limits a and

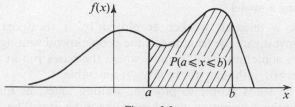

Figure 9.9

b and this is given by $\int_a^b f(x)dx$. So we have $P(a \leqslant x \leqslant b) = \int_a^b f(x)dx$. Notice that this expression will also equal $P(a < x \leqslant b)$ and $P(a < x < b)$ and $P(a \leqslant x < b)$ since $P(x = a) = P(x = b) = 0$.

This tallies very well with the result we obtained in the discrete case. Suppose x is a random variable taking integer values only, then we have

for instance $P(0 \leqslant x \leqslant 3) = P(0) + P(1) + P(2) + P(3) = \sum_{x=0}^{3} P(x)$. In the

continuous case however $P(0 \leqslant x \leqslant 3) = \int_0^3 f(x)dx$. The formulae correspond as we can replace the summation by an integral.

We can see that the integral formula does work in the rather trivial case where the distribution is uniform over the range 0–10 and hence $f(x) = \frac{1}{10}$, $0 \leqslant x \leqslant 10$; for $P(0 \leqslant x \leqslant 3) = \int_0^3 \frac{1}{10}dx = \frac{1}{10}\int_0^3 dx = \frac{3}{10}$, which is what we would expect.

Can any function $f(x)$ represent some possible probability density function? It is clear the $f(x)$ must be positive for all x, for otherwise $f(x_0).\delta x$, which is the approximate probability that x lies between x_0 and $x_0 + \delta x$, would be negative for that value x_0 for which $f(x_0)$ is negative.

A further condition is analogous to the condition for the discrete case that a set of positive numbers $p_0, p_1, p_2 \ldots$ could correspond to a possible probability distribution provided $\Sigma p = 1$. Here again the total probability that x takes one of its possible values must be 1. We can write this as

$$\int_{all\ x} f(x)dx = 1 \quad \text{or as} \quad \int_{-\infty}^{\infty} f(x)dx = 1,$$

meaning that the integral of $f(x)$ over the whole range of x must be 1.

To gather up all the results we have had so far, we find that if a random variable x can take a continuous range of values, then we need to introduce a new function, the **probability density function** $f(x)$, which tells us the "height" of the curve at any point x. We have

(a) the probability of any particular value of x is 0,

(b) $P(a \leqslant x \leqslant b) = \int_a^b f(x)dx$,

(c) $\int_{-\infty}^{\infty} f(x)dx = 1$, and $f(x) \geqslant 0$ for all x.

9.3. Choosing a model

Suppose a research worker employed by a transport company decides to investigate the lengths of time people spend waiting for buses, and chooses some (mythical?) route where the buses run at about ten-minute intervals, although not to any set timetable.

Before he sends people to plot the waiting times, he might try to predict what the distribution would look like. First, it will be continuous rather than discrete, since the time spent by a person in the bus queue can take any value in a continuous range. Of course the person who actually has the job of measuring the waiting times will not be able to plot a continuous curve; he will only be able to measure to a certain accuracy, say 5 seconds, and will only be able to find the number of waiting times within 5-second ranges of time.

(This is in fact true of any experimental data; we cannot measure anything which varies on a continuous scale, for example length, time or weight, to more than a certain number of decimal places. Thus all data are, in a sense, discrete. However, for the purpose of a theoretical model, it is generally easier to disregard this and work out the results as if our measuring was perfect, since continuous curves are easier to deal with than discrete ones with a large number of values.)

If the research worker first assumes that the buses always run at exactly ten-minute intervals and that everyone is able to get on the first bus, then there would be no waiting times longer than 10 minutes. Also, since people have no set timetable to tell them when the next bus will arrive, he might reasonably expect them to join the queue at random during the ten-minute intervals. So, in this case he would expect the waiting times to be uniformly distributed in the interval 0–10 minutes. This would give him exactly the same situation as we had before in considering the distribution of random numbers in the range 0–10.

Let us work out the mean and standard deviation of this distribution.

As we saw before the mean is fairly obvious as it must by symmetry be 5 minutes; but if the distribution was not uniform, how would we find it?

If we look at the analogy with the discrete case, we had there $E(x) = \Sigma x P(x)$. Since now $P(x) = 0$ for all values of x, we clearly cannot use this formula as it stands. Suppose we split our range of 10 minutes into a large number of small intervals, each of length δx. Then since $P(x_0 \leqslant x \leqslant x_0 + \delta x) \simeq f(x_0) \, \delta x$, and, at least approximately, x will have value x_0 over this small interval, the mean of x will be almost equal to

$$\sum_{\text{all } x_0} x_0 f(x_0) \delta x.$$

If we now take the limit as $\delta x \to 0$ we get $E(x) = \int_{-\infty}^{\infty} xf(x)dx$.

So there is again a close parallel between the discrete and continuous formulae

$$E(x) = \Sigma x P(x) \longleftrightarrow E(x) = \int_{-\infty}^{\infty} xf(x)dx$$

Exactly the same reasoning works with the variance,

$$\text{Var}(x) = E\big[(x - E(x))^2\big] = E(x^2) - [E(x)]^2.$$

$$E\big[(x - E(x))^2\big] = \Sigma(x - E(x))^2 P(x) \longleftrightarrow E[(x - E(x))^2]$$
$$= \int_{-\infty}^{\infty} (x - E(x))^2 f(x)dx$$

$$E(x^2) = \Sigma x^2 P(x) \longleftrightarrow E(x^2) = \int_{-\infty}^{\infty} x^2 f(x)dx.$$

Thus $\text{Var}(x) = \int_{-\infty}^{\infty} [x - E(x)]^2 f(x)dx = \int_{-\infty}^{\infty} x^2 f(x)dx - [\int_{-\infty}^{\infty} xf(x)dx]^2.$

So we can work out the mean and standard deviation of the bus times without more ado since $f(x) = \frac{1}{10}, 0 \leqslant x \leqslant 10$.

$$E(x) = \int_0^{10} xf(x)dx = \int_0^{10} \tfrac{1}{10}x \, dx = \tfrac{1}{10}\big[\tfrac{1}{2}x^2\big]_0^{10} = 5 \text{ as we expected.}$$

$$E(x^2) = \int_0^{10} x^2 f(x)dx = \int_0^{10} \tfrac{1}{10}x^2 dx = \tfrac{1}{10}\big[\tfrac{1}{3}x^3\big]_0^{10} = \tfrac{100}{3}$$

$$\text{Var}(x) = \tfrac{100}{3} - 25 = \tfrac{25}{3}. \quad \sigma = \sqrt{\tfrac{25}{3}} \simeq 2 \cdot 9.$$

Thus if he were to do a survey of such a perfect bus service, he would expect the mean waiting time to be 5 minutes with standard deviation 2·9 minutes.

Suppose he found that this model did not fit very well with the real situation and that, although the interval between buses was about ten minutes, it was occasionally more. He would have to alter his model to allow for this. Suppose he still assumes that the distribution is uniform between 0 and 10, which means that people arrive randomly in the ten minutes after the last bus has left, so that the distribution still has the basic shape of Fig. 9.10.

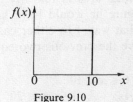

Figure 9.10

However he now has to allow for possible waiting times longer than ten minutes. He reasons that, if x is greater than 10, the larger it is the less likely it is that the wait will be that long. In other words the probability of waiting between 12 and 13 minutes will be greater than the probability of waiting between 20 and 21 minutes. It looks as though he will have a picture of one of the forms in Figs. 9.11–9.13.

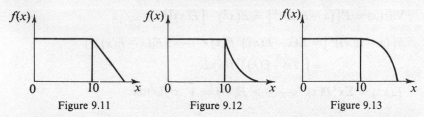

Figure 9.11   Figure 9.12   Figure 9.13

Which of these he chooses would depend ultimately on which best fitted the experimental data. Suppose he chooses the simplest, Fig. 9.11, and assumes in addition that no person waits longer than 20 minutes for a bus. Then $f(x)$ must have the form:

$$f(x) = k, 0 \leqslant x \leqslant 10$$
$$= ax+b, 10 \leqslant x \leqslant 20$$
$$= 0 \text{ otherwise.}$$

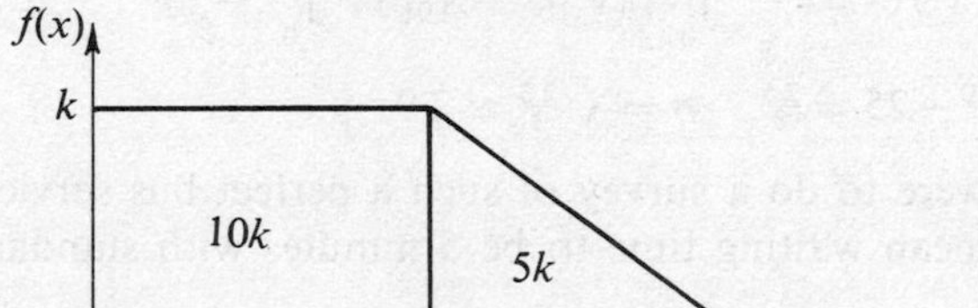

Figure 9.14

He has to find the three constants k, a and b. He knows that

(i) $f(x)$ must be continuous at $x = 10$ which tells him that $k = 10a+b$;
(ii) $f(x) = 0$ at $x = 20$ which gives him $20a+b = 0$;
(iii) the total area under the curve is 1.

From the diagram the total area is $15k$. (If the area had not been so easy to see from the figure he would have had to explicitly integrate $f(x)$ and find the area that way.) In either case he has $15k = 1$ and so $k = \frac{1}{15}$. He can now solve the previous two equations to find $a = -\frac{1}{150}$ and $b = \frac{2}{15}$. So, he has

$$f(x) = \frac{1}{15}, 0 \leqslant x \leqslant 10$$
$$= -\frac{1}{150}x+\frac{2}{15}, 10 \leqslant x \leqslant 20.$$

If he now wanted to find the mean of the distribution he would have

$$E(x) = \int_0^{20} xf(x)dx = \int_0^{10} \tfrac{1}{15}x\ dx + \int_{10}^{20} x(-\tfrac{1}{150}x+\tfrac{2}{15})dx$$

$$= \tfrac{1}{15}\left[\tfrac{1}{2}x^2\right]_0^{10} - \tfrac{1}{150}\left[\tfrac{1}{3}x^3\right]_{10}^{20} + \tfrac{2}{15}\left[\tfrac{1}{2}x^2\right]_{10}^{20} = \tfrac{70}{9}$$

So the mean waiting time for a bus given by the model is roughly 7·8 minutes. The research worker could check whether this theoretical model was a suitable one by finding the mean of the readings he obtains and judging whether the predicted mean is close enough to it.

9.4. The cumulative distribution function

In this second model of the waiting times for buses we might well be interested in the probability of waiting less than ten minutes for a bus. This will be given by

$$P(0 \leqslant x \leqslant 10) = \int_0^{10} f(x)dx = \int_0^{10} \tfrac{1}{15}dx = \tfrac{2}{3},$$

and with our usual notation we could denote this by $F(10)$ where $F(a)$, the **cumulative distribution function** (c.d.f.), is defined as usual to be the probability that x is less than or equal to a.

For a general probability density function we will have

$$F(a) = \int_{-\infty}^{a} f(x)dx$$

(where the integral is taken over all values of x less than or equal to a). We notice again the analogy between this and the discrete case where $F(a) = \sum_{x \leqslant a} P(x)$.

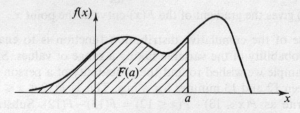

Figure 9.15

Let us find the general formula for F for the buses example. We have $f(x) = \tfrac{1}{15}, 0 \leqslant x \leqslant 10, f(x) = -\tfrac{1}{150}x+\tfrac{2}{15}, 10 \leqslant x \leqslant 20$. The way we evaluate the integral for F will depend upon the value of a. If $a \leqslant 0$ clearly $F(a) = 0$.

If $0 \leqslant a \leqslant 10$ we have $F(a) = \int_0^a \frac{1}{15}dx = \frac{1}{15}a$,

$10 \leqslant a \leqslant 20$ we have $F(a) = \int_0^{10} \frac{1}{15}dx + \int_{10}^a [-\frac{1}{150}x + \frac{2}{15}]dx$

$$= \frac{2}{3} - \frac{1}{150} \times \frac{1}{2}(a^2 - 10^2) + \frac{2}{15}(a - 10) = \frac{2}{15}a - \frac{1}{300}a^2 - \frac{1}{3}.$$

If $a \geqslant 20$ clearly $F(a) = 1$.

Writing x instead of a we have

$$\begin{aligned} F(x) &= 0, & x &\leqslant 0 \\ &= \frac{1}{15}x, & 0 &\leqslant x \leqslant 10 \\ &= \frac{2}{15}x - \frac{1}{300}x^2 - \frac{1}{3} & 10 &\leqslant x \leqslant 20 \\ &= 1 & x &\geqslant 20 \end{aligned}$$

If we plot the function $F(x)$ we obtain Fig. 9.16.

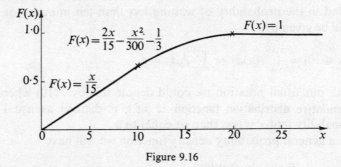

Figure 9.16

Notice that $F(x)$ is continuous at $x = 0$ and $x = 10$ and $x = 20$.

We can also express the relationship between $f(x)$ and $F(x)$ the other way round. Since

$$F(a) = \int_{-\infty}^a f(x)dx \text{ we must have } f(x) = \frac{d}{dx}[F(x)]$$

so that $f(x)$ gives the gradient of the $F(x)$-curve at the point x.

One use of the cumulative distribution function is to enable us to find the probability of the variable taking a range of values. Suppose in the bus example we wished to find the probability of a person having to wait between 12 and 13 minutes for a bus. This is $P(12 \leqslant x \leqslant 13)$ which we can write as $P(x \leqslant 13) - P(x \leqslant 12) = F(13) - F(12)$. Substituting 12 and 13 in our formula for F above we can get the result. In general we will have $P(a \leqslant x \leqslant b) = F(b) - F(a)$.

We can also use F to find the median and quartiles of the distribution. The median M will be that value for which $F(M) = \frac{1}{2}$ and the two quartiles Q_1 and Q_3 the values for which $F(Q_1) = \frac{1}{4}$ and $F(Q_3) = \frac{3}{4}$ respectively.

EXERCISE 9A

1. A probability density function $f(x)$ has the form shown:

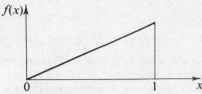

   (i) Write down an expression for $f(x)$,
   (ii) find and sketch $F(x)$,
   (iii) calculate the mean and the median,
   (iv) find $P(x \leqslant \frac{1}{2})$.
2. A random variable x is equally likely to take any value between a and b, $(a < b)$ (the general *uniform distribution*). Find $f(x)$, $F(x)$, $E(x)$ and Var (x).
3. For the graph shown find $f(x)$, $F(x)$ and Var(x).

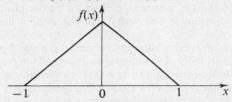

4. Find the standard deviation of the waiting times for buses with the second model used in the chapter.
5. A garage is supplied with petrol once a week. The weekly demand, x thousand litres, has the continuous probability density function $f(x) = 5(1-x)^4$, $0 \leqslant x \leqslant 1$. What must be the capacity of the petrol tank if the probability that it will be exhausted in any given week is to be only 0·01? (J.M.B.)
6. The continuous random variable x has mean μ and variance σ^2. Show that $E(ax+b) = a\mu+b$ and Var$(ax+b) = a^2\sigma^2$ for any constants a and b.
7. A variate x can assume values only between 0 and 5, and the equation of its frequency curve is $y = A \sin \frac{1}{5}\pi x$, $(0 \leqslant x \leqslant 5)$, where A is a constant such that the area under the curve is unity. Determine the value of A and obtain the median and quartiles of the distribution. Show also that the variance of the distribution is $50 \left\{ \dfrac{1}{8} - \dfrac{1}{\pi^2} \right\}$. (J.M.B.)
8. A random variable x has cumulative density function $F(x) = \begin{cases} 0 & x \leqslant 0 \\ kx^3 & 0 < x \leqslant 3; \\ 1 & x > 3 \end{cases}$ find and sketch $f(x)$ and find σ^2.
9. Given the continuous frequency function $f(x) = 2/x^2$ where $1 \leqslant x \leqslant 2$, determine the mean and variance of x and find the probability that x exceeds 1·5. Calculate also the median and quartile values for x and state the interquartile range. (J.M.B.)
10. A variate x can assume values only between 0 and a, and the equation of its frequency curve is $y = A(a-x)^2$, $(0 \leqslant x \leqslant a)$, where A and a are constants such that the area under the curve and the mean of the distribution are both unity. Determine the numerical values of A and a, and find the 10th and 90th percentiles of the distribution. (J.M.B.)

11.

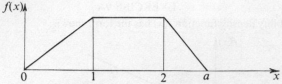

For $f(x)$ shown, if the mean is $\frac{9}{8}$, find a, $f(x)$, and the probability that the variable lies between 1·5 and 2·5.

12. A probability density function of a random variable x is defined as follows:

$$f(x) = \begin{cases} x(x-1)(x-2) & \text{for } 0 \leqslant x \leqslant 1 \\ \lambda & \text{for } 1 < x \leqslant 3 \\ 0 & \text{otherwise} \end{cases}$$

where λ is a suitable constant. Calculate the expectation μ of x. What is the probability that x is less than or equal to μ? (M.E.I.)

13. A continuous probability distribution has frequency density $f(x)$ where $f(x) = a + bx + cx^2$ between $x = 0$ and $x = 1$, and $f(x) = 0$ outside this range. The mean is $\frac{2}{3}$ and the variance $\frac{1}{45}$. Find the values of a, b, and c. Find the mode of the distribution and verify that the median lies between 0·65 and 0·66. (M.E.I.)

14 It is known that an event must occur between 2 pm and 5 pm on a certain day, and that the probability of it occurring in any small interval of time is inversely proportional to the square of the time that has elapsed since noon on that day. Find by what time it is an even chance that the event has occurred. Find also the probability of the event occurring between 4 pm and 5 pm. (J.M.B.)

15. Jean and Jane agree to meet at a certain cafe for tea one day. They both choose their time of arrival at random between 4 pm and 5 pm and each waits 10 minutes. What is the probability that they meet?

16. A chord of a circle of radius a is drawn at random parallel to a given straight line, all distances from the centre of the circle being equally probable. Show that the expected value of the length of the chord is $\frac{1}{2}\pi a$ and the expected value of the square of the length is $\frac{8}{3}a^2$. Hence determine the standard deviation of this continuous distribution. (M.E.I.)

17. Show that:

$$\int_0^4 x^{3+m}(4-x)^2 dx = \frac{2.4^{6+m}}{(4+m)(5+m)(6+m)} \text{ for } m = 0, 1, 2.$$

Hence given that a variable x has p.d.f. $f(x)$ where $f(x) = kx^3(4-x)^2$, $0 \leqslant x \leqslant 4$, find k, $E(x)$ and $\text{Var}(x)$.

18. If a variable x with cumulative distribution function $F(x)$ can take only values between a and b, where a and b are finite and $a < b$, show, by integration by parts, that:

$$E(x) = b - \int_a^b F(x)dx.$$

19. A point P is taken at random in a line AB of length a, all positions of the point being equally likely. Find the expected value of the area of the rectangle AP.PB, and show that the probability that the area exceeds $a^2/8$ is $1/\sqrt{2}$. (J.M.B.)

20. A probability density $p(x)$ is given by $p(x) = Cx(4-x)$, $(0 \leqslant x \leqslant 4)$, where C is a positive constant. Find the value of C and sketch the distribution. Explain why the standard deviation of this distribution is the same as the standard deviation of the distribution with probability density $p(x) = C(4-x^2)$, ($|x| \leqslant 2$). For the first distribution, calculate the mean, the standard deviation and the mode. Calculate also the probability that a value of the variable x taken at random will lie more than one standard deviation away from the mean.

(M.E.I.)

9.5. The distribution of a function of a variable and the exponential distribution

In the last chapter we found that, under certain assumptions, the number of flaws in a length of cotton woven on a loom had a Poisson distribution. Suppose we are now interested in the distribution of the length of material which is free from flaws. Let the first flaw occur after a length x (measured from any convenient point). We will find $F(a) = P(x \leqslant a)$, which can be thought of as the probability that there is at least one flaw in a piece of material of length a. In the particular example we considered there was on average 1 fault in every 10 metres of cloth, so in a piece of length a metres there will be an average of $a/10$ faults. The number of faults in this piece of material will thus have the $\mathscr{P}(a/10)$ distribution and in particular the probability that there will be no faults in the piece will be $e^{-a/10}$. So the probability that there will be at least one fault is $1 - e^{-a/10}$, which gives $F(a) = 1 - e^{-a/10}$. Or, writing x instead of a, $F(x) = 1 - e^{-x/10}$.

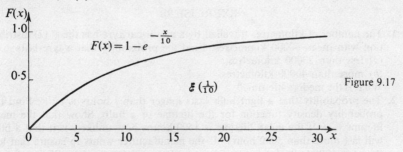

Figure 9.17

We can now differentiate this and obtain $f(x) = \dfrac{d}{dx}[F(x)] = \tfrac{1}{10}e^{-x/10}$.

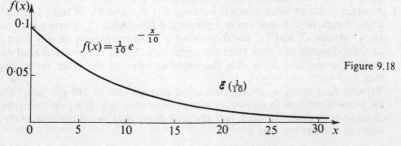

Figure 9.18

We have shown that if the average number of faults is 1 per 10 metres, then the length of material free from faults has a probability density function $f(x) = \frac{1}{10}e^{-x/10}$, $x \geqslant 0$. In general if the average number of faults per metre were λ the probability density function would be $\lambda e^{-\lambda x}$, $x \geqslant 0$. This is known as the **exponential distribution** and is denoted by $\mathscr{E}(\lambda)$. It is our first example of a continuous distribution with an infinite range. The mean of the distribution is given by

$$E(x) = \int_0^\infty xf(x)dx = \int_0^\infty \lambda x e^{-\lambda x}dx = \left[-xe^{-\lambda x}\right]_0^\infty + \int_0^\infty e^{-\lambda x}dx$$

integrating by parts. The first term is zero, so

$$E(x) = \int_0^\infty e^{-\lambda x}dx = \left[-(1/\lambda)e^{-\lambda x}\right]_0^\infty = 1/\lambda$$

(which incidentally verifies that

$$\int_{-\infty}^\infty f(x)dx = \int_0^\infty \lambda e^{-\lambda x}dx = 1).$$

So the average length of cloth free from faults is $1/\lambda$ metre.

This method of first finding $F(x)$ and then differentiating it to find $f(x)$ is an extremely useful one for finding the p.d.f. of one variable from that of a different related variable as you will see in the following Exercise.

EXERCISE 9B

1. The number of kilometres travelled by a motor-car tyre has the $\mathscr{E}(\lambda)$ distribution with mean 30000 kilometres. Find the probability that a tyre lasts:
   (a) less than 20000 kilometres,
   (b) more than 40000 kilometres.
   What is the median life-time?
2. The probability that a light bulb lasts longer than t hours is $e^{-t/\mu}$. Find the probability density function for the lifetime of a bulb. Show that the mean lifetime is μ. If the mean lifetime is 1500 hours, how unlikely is it that a bulb will last more than 3000 hours? If the manufacturer wants to ensure that less than 1 in 1000 bulbs fail before 5 hours, what is the lowest mean lifetime he can allow his bulbs to have? (S.M.P.)
3. Find the variance and interquartile range of $\mathscr{E}(\lambda)$.
4. A target contains three consecutive rings C_1, C_2, and C_3 of radii 1, 2 and 3 units respectively. A shot scores 3 points if it falls inside C_1, 2 between C_1 and C_2, 1 between C_2 and C_3, and 0 outside C_3. If the probability density function of r, the distance of a shot from the centre, is e^{-r}, find the probability of each possible score, and show that the expected value of the score for a single shot is $3 - e^{-1} - e^{-2} - e^{-3}$. (J.M.B.)
5. Vehicles pass along a certain road at an average rate of 180 per hour. Find the p.d.f. of the time in minutes between consecutive cars. If a pedestrian needs 20 seconds to cross the road, find the probability that he will be able to do so between the next two cars.

6. For a certain industrial firm it is found that if a person joins the firm the probability that he is still there at time x later, where x is measured in years, is $\frac{1}{3}e^{-\frac{1}{4}x}+\frac{2}{3}e^{-\frac{1}{4}x}$. Find the probability that a person spends between 2 and 4 years with the firm, and the mean and variance of the length of time he works there.

7. A random variable X is uniformly distributed over the interval $(-1, 1)$. Find $P(X^2 \leqslant x)$ for any given value of x and hence find the probability density function of the random variable X^2. Find the mean and variance of X^2 from this p.d.f. How could you have found them without first finding the p.d.f.?

8. The average speeds of a number of vehicles each making the same journey of 64 km may be assumed to be rectangularly distributed over the interval 32 km/h to 64 km/h. Find the probability distribution of the time T taken to make the journey. For what proportion of vehicles will T lie between 1 h 30 min and 1 h 45 min? (J.M.B.)

9. A lighthouse is situated at a distance l from a straight coastline and sends out a continuous narrow beam of light simultaneously in opposite directions. The light source is rotating at a constant angular velocity. At an arbitrary instant of time it illuminates the coastline at a point X. The direction of the light beam at this instant makes an angle θ with the line from the lighthouse to the nearest point on the coastline O. With respect to O as origin, the point X has coordinate x. Assuming that θ is uniformly distributed over the range $-\frac{1}{2}\pi$ to $\frac{1}{2}\pi$, show that x has probability density function

$$f(x) = \frac{1}{\pi l} \cdot \frac{1}{1+(x/l)^2} \quad (-\infty < x < \infty).$$

Sketch this distribution. Find the probability that, at an arbitrary instant of time, the light beam strikes the coastline at a distance from O greater than $2l$. (J.M.B.)

10. A point P is chosen at random on a fixed radius of a circle of radius a so that its distance x from the centre of the circle has a rectangular distribution on the interval $0 \leqslant x \leqslant a$. Find the probability density function of X, the length of the chord through P that is perpendicular to the fixed radius. Determine the probability that X is less than a. (J.M.B.)

11. In Exercise 9A, question 15, if variable y denotes the time in minutes after 4 pm at which the last to arrive reaches the cafe, write down $P(y \leqslant a)$ for any a, $0 \leqslant a \leqslant 60$, and hence find and sketch the p.d.f. of y. Find the mean and variance of y. Carry out similar calculations for the time at which the first one arrives.

The Normal Distribution

10.1. The shape of the distribution

Suppose we told 100 people each to go and measure roughly the length of a field, or each to throw a dart aiming for the centre of the board and measure the height at which it lands, or each to pour out roughly half a litre of milk and then measure the actual quantity they had poured, what would the distribution curves look like?

If you measured the lengths of 100 cows, or laurel leaves, or the mathematical attainment of 100 children, what would be the shape of the distribution curves?

The answers to these questions is "much the same in both sets of examples", as such apparently diverse situations have at least these three things in common:

(a) The variables being measured are all continuous.

(b) The distributions are likely to be symmetric about the mean; for example, it seems as probable that an estimate of the field length will be between 1 and 2 m too small as between 1 and 2 m too large.

(c) The probability of getting a result in a certain interval will decrease with the distance of the interval from the mean; for example, it is more likely that someone's measurement of the field length will be between 1 and 2 m too low than that it will be between 8 and 9 m too low.

These three observations point to a distribution which looks like Fig. 10.1.

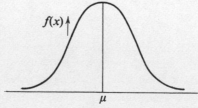

Figure 10.1

132

It shows that we have allowed the variable to take all possible values. When we apply it to some of the variables in the second set of examples, this seems odd, as it allows the possibility of negative lengths. However, in these situations the mean will be positive and the distribution will slope off very steeply in its "tails" in such a way that the probability of obtaining a negative length is negligible. The curve can therefore still act as a serviceable model of such examples.

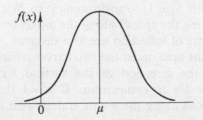

Figure 10.2

Such a curve was in fact comprehensively studied by Gauss in connection with the distribution of errors, and is therefore often known as the *Gaussian* or *error* curve, although we shall prefer to give it its more usual name of the **normal** curve. As well as situations involving random errors, it is extremely useful in the description of measurable biological characteristics, and we will later see other examples of its applications.

Not all the examples we have discussed will be fitted by exactly the same distribution curve; they will differ firstly in their mean μ. The mean as we have seen lies on the line of symmetry of the curve, and is equal to both the mode and the median. The variables will differ also in their variance σ^2; those with a larger value of the variance will have a distribution which is more spread out about the mean than those with a smaller variance. In fact, we shall be able to prove later that the two symmetric points of inflection, where the numerical value of the gradient is a maximum (and therefore $f''(x) = 0$) occur at a distance σ on either side of the mean.

We will see below that the shape of any particular normal curve is determined by the values of the parameters μ and σ^2, so that we shall refer to the normal distribution with mean μ and variance σ^2 as $N(\mu, \sigma^2)$.

We can see that the curve $N(3, \sigma^2)$ will be identical to $N(2, \sigma^2)$ except that it is translated 1 unit along the x-axis.

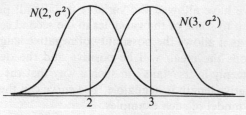

Figure 10.3

Also, if we start with $N(\mu, 1)$, transforming it to $N(\mu, \sigma^2)$ where $\sigma^2 > 1$, will involve increasing the spread along the x-axis by a factor σ in such a way that the points of inflection are at a distance $\pm\sigma$ from the mean instead of ± 1. As the area under the two curves remains constant at one unit, distances in the direction of the vertical $f(x)$-axis will be decreased by a factor $1/\sigma$ to compensate. If $\sigma^2 < 1$ the situation will be reversed as the spread along the x-axis is decreased and corresponding values along the $f(x)$-axis increased.

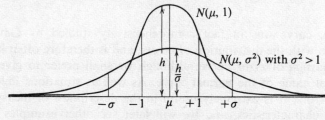

Figure 10.4

We can therefore transform any normal curve into any other by a combination of translation and stretching (or shrinking) parallel to the axes. Because of this, we shall try to find the mathematical equation for the probability density function first for the special case $N(0, 1)$ and then transform it to obtain the general equation.

10.2. The probability density function

We can see that the probability density function for $N(0, 1)$ must satisfy the following conditions:
(1) $f(x)$ is symmetrical about $x = 0$, i.e. $f(-x) = f(x)$
(2) $f(x) \geqslant 0$ for all x
(3) $f(x) \to 0$ as $x \to \pm\infty$
(4) $f(0)$ is finite
(5) $f(x)$ has a maximum at $x = 0$

(6) the area under the curve is 1, i.e. $\int_{-\infty}^{\infty} f(x)dx = 1$

(7) the variance is 1, i.e. since $\mu = 0$, $\int_{-\infty}^{\infty} x^2 f(x)dx = 1$.

As the curve appears to be completely smooth, we will try to find a single function of x which satisfies all these conditions.

Condition 1 tells us that $f(x)$ must be a function of even powers of x only, i.e. x^2, x^4, etc. But $f(x) = x^2$ or $f(x) = x^4$ fail to satisfy condition 3 and $f(x) = 1/x^2$ or $f(x) = 1/x^4$ though satisfying condition 3 do not fit condition 4.

The next most obvious function to try might be a power function, e.g. $f(x) = a^{x^2}$. This satisfies conditions 1, 2 and 4 at least, provided $a > 0$. In order to make $f(x)$ easier to integrate and differentiate we will choose to write it as $f(x) = e^{kx^2}$ (which means choosing k so $e^k = a$), where $e = 2.718\ldots$ is the irrational number we met first when studying the Poisson distribution. In order to satisfy condition 3 we must have k negative.

So $f(x) = e^{kx^2}$ satisfies conditions 1–4 provided $k < 0$, and more generally so will $f(x) = Ae^{kx^2}$ provided also $A > 0$. Since $\dfrac{d}{dx}e^x = e^x$ we have $f'(x) = 2kxAe^{kx^2}$ so that $f'(0) = 0$ and hence $f(x)$ satisfies condition 5 as well. (It is obvious that this turning point is indeed a maximum.) Now all we need to check is the two conditions 6 and 7 involving integrals of $f(x)$.

Taking condition 7 first we must have $\int_{-\infty}^{\infty} Ax^2 e^{kx^2}dx = 1$

$\int_{-\infty}^{\infty} Ax^2 e^{kx^2}dx = \left[Axe^{kx^2}/2k\right]_{-\infty}^{\infty} - \int_{-\infty}^{\infty}(Ae^{kx^2}/2k)dx$, integrating by parts.

Since $xe^{kx^2} \to 0$ as $x \to \pm\infty$, the first term is zero, and since $\int_{-\infty}^{\infty} Ae^{kx^2}dx = 1$ if condition 6 is to be satisfied, we must have $-1/2k = 1$ and hence $f(x) = Ae^{-\frac{1}{2}x^2}$.

Condition 6 is much harder to investigate as $\int_{-\infty}^{\infty} e^{-\frac{1}{2}x^2}dx$ is extremely difficult to evaluate directly. However it is not too difficult to use Simpson's rule or one of the other numerical methods of integration for the region, say $-4 \leqslant x \leqslant 4$, in order to verify partially that

$$\int_{-\infty}^{\infty} e^{-\frac{1}{2}x^2}dx = \sqrt{(2\pi)}$$

and this can be proved theoretically. In order to satisfy condition 6 we must therefore have $A = 1/\sqrt{(2\pi)}$. We therefore have all our conditions satisfied by

$$f(x) = \frac{1}{\sqrt{(2\pi)}}e^{-\frac{1}{2}x^2}$$

as the probability density function for $N(0, 1)$. (Sometimes you will see $\phi(x)$ written instead of $f(x)$ for this particular probability density function.)

If we now transform this to the case $N(0, \sigma^2)$ we must stretch each x by a factor σ, $x \to X$ where $X = \sigma x$, i.e. $x = X/\sigma$. Also in order to keep the area constant we have to shrink the height $f(x)$ by a factor $1/\sigma$. Hence the new equation must be

$$f(X) = \frac{1}{\sigma\sqrt{(2\pi)}} e^{-\frac{1}{2}X^2/\sigma^2},$$

or, rewriting it in terms of x,

$$f(x) = \frac{1}{\sigma\sqrt{(2\pi)}} e^{-\frac{1}{2}x^2/\sigma^2} \text{ is the equation of } N(0, \sigma^2).$$

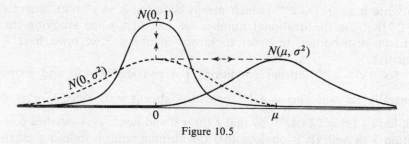

Figure 10.5

If we now extend this to the case $N(\mu, \sigma^2)$, all we have to do is to translate the distribution along the x-axis so that $x \to u$ where $u = x + \mu$, i.e. $x = u - \mu$. Substituting we have

$$f(u) = \frac{1}{\sigma\sqrt{(2\pi)}} e^{-\frac{1}{2}(u-\mu)^2/\sigma^2},$$

or, rewriting in terms of x,

$$f(x) = \frac{1}{\sigma\sqrt{(2\pi)}} e^{-\frac{1}{2}(x-\mu)^2/\sigma^2} \text{ as the equation of } N(\mu, \sigma^2).$$

We have therefore reached the conclusion that the negative exponential function

$$f(x) = \frac{1}{\sigma\sqrt{(2\pi)}} e^{-\frac{1}{2}(x-\mu)^2/\sigma^2}$$

has a curve with roughly the shape we would need in any model which would fit all the examples given at the beginning of the chapter. This does not guarantee that it will always be the curve giving the best fit in any particular case. However whenever the variations in a measurement can be assumed to be due to a large number of sources of random error, it can be shown theoretically that its distribution will be at least approximately of this form.

10.3. Transformation from $N(\mu, \sigma^2)$ to $N(0, 1)$

Because the equation for $N(0, 1)$ is so much simpler than that for $N(\mu, \sigma^2)$ and because a simple transformation will enable any normal curve to be transformed to it, $N(0, 1)$ is known as the **standard normal distribution**, and we call the process of transforming $N(\mu, \sigma^2)$ to $N(0, 1)$ **standardizing**.

Suppose we start with $N(\mu, \sigma^2)$.

$N(\mu, \sigma^2) \rightarrow N(0, \sigma^2)$: $x \rightarrow v$ where $v = x - \mu$.

Then $N(0, \sigma^2) \rightarrow N(0, 1)$: $v \rightarrow z$ where $z = v/\sigma$

so $N(\mu, \sigma^2) \rightarrow N(0, \sigma^2) \rightarrow N(0, 1)$: $x \rightarrow v \rightarrow z$ where $z = v/\sigma = (x - \mu)/\sigma$.

If the random variable x is distributed as $N(\mu, \sigma^2)$, then the random variable z is distributed as $N(0, 1)$ where $z = (x - \mu)/\sigma$. (We shall call $(x - \mu)/\sigma$ the **standardized normal variate** and always refer to it as z.)

This transformation is frequently used to compare examination marks; for instance you might want to find out whether a mark of 73 in a rather easy English examination where the mean mark was 65 and the standard deviation 5 was really any better than a mark of 68 in a difficult mathematics paper where the mean was only 50 but the standard deviation was 10. We can see that in the English case $z = (73 - 65)/5 = 1 \cdot 6$ so that the mark obtained was $1 \cdot 6$ standard deviations above the mean, whereas in the mathematical case $z = (68 - 50)/10 = 1 \cdot 8$ and so the mark was $1 \cdot 8$ standard deviations above its mean. Here therefore we would consider the 68 for mathematics as the better mark.

10.4. The cumulative distribution function

A zoologist has, from a sample of a certain kind of tropical snake, found that their mean length fully grown is 750 mm with a standard deviation of 150 mm. Is it likely that there are many snakes more than 1 m (1000 mm) long?

As it stands we cannot really answer this question, as it may be physically impossible for the species to grow to more than 1 m, but if we assume that the distribution of the snake lengths fits approximately the normal model, then we can use our knowledge of the distribution to estimate the proportion. We have

$$P(x > 1000) = 1 - P(x \leqslant 1000) = 1 - F(1000)$$

where $F(x)$ is the cumulative distribution function. But

$F(a) = \int_{-\infty}^{a} f(x)dx$, so by definition $F(a) = \int_{-\infty}^{a} \frac{1}{\sqrt{2\pi}\,\sigma} e^{-\frac{1}{2}(x-\mu)^2/\sigma^2}$ for $N(\mu, \sigma^2)$.

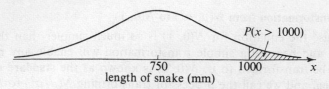

Figure 10.6

This looks rather formidable, so we will try standardizing the variable, using the transformation $z = (x - \mu)/\sigma$, or in this case $z = (x - 750)/150$. The value of z which corresponds to $x = 1000$ is $z = (1000 - 750)/150 = \frac{5}{3}$; also if $x > 1000$ we have $z > \frac{5}{3}$,

$$\text{so } P(x > 1000) = P(z > \tfrac{5}{3}) = 1 - P(z \leqslant \tfrac{5}{3}) = 1 - \Phi(\tfrac{5}{3}),$$

where $\Phi(z) = \int_{-\infty}^{z} e^{-\frac{1}{2}x^2} dx.$

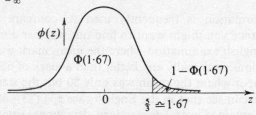

Figure 10.7

Φ is then the cumulative distribution function for the standard normal distribution $N(0, 1)$.

We saw before that $\int_{-\infty}^{\infty} e^{-\frac{1}{2}x^2} dx$ was difficult to evaluate, and we have even worse problems with $\int_{-\infty}^{z} e^{-\frac{1}{2}x^2} dx$ for a general value z as this integral does not have a simple formula. The best we can do is to evaluate the integral by numerical means and print a table of values of $\Phi(z)$ for a variety of values of z. In fact as $N(0, 1)$ is symmetric about $z = 0$, it is only necessary to print the table for positive values of z. We can see from the diagram that $\Phi(-z) = 1 - \Phi(z)$.

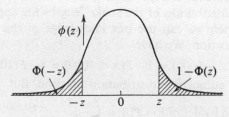

Figure 10.8

We can now see the reason for converting any normal variable with distribution $N(\mu, \sigma^2)$ to a standardized variable with $N(0, 1)$, as we now need only the table of values of $\Phi(z)$ instead of tables of $F(x)$ for many differing values of μ and σ^2. The table of Φ is on page 242. It is clear that $\Phi(0) = 0.5$ and hence $\Phi(z) > 0.5$ for $z > 0$.

Returning to our problem we have $P(x > 1000) = 1 - \Phi(\frac{5}{3})$. From the tables we have $\Phi(\frac{5}{3}) \simeq 0.952$. Hence $P(x > 1000) \simeq 1 - 0.952 = 0.048$. Our answer therefore is that we should expect about 4.8% of the snakes to have lengths greater than 1 m.

What proportion of them would we expect to have lengths between 450 and 900 m?

We have $P(450 \leqslant x \leqslant 900) = P(x \leqslant 900) - P(x \leqslant 450) = \Phi(b) - \Phi(a)$ where $b = (900 - 750)/150 = 1$ and $a = (450 - 750)/150 = -2$.

$P(450 \leqslant x \leqslant 900)$

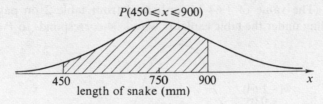

length of snake (mm)

Figure 10.9

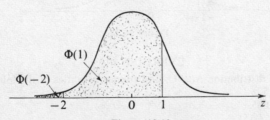

Figure 10.10

From the tables we have $\Phi(1) \simeq 0.8413$
and $\Phi(-2) = 1 - \Phi(2) \simeq 1 - 0.9772 = 0.0228$.
So $P(450 \leqslant x \leqslant 900) \simeq 0.8413 - 0.0228 = 0.8185$, and
therefore we would expect roughly 82% of the snakes to have lengths between 450 and 900 mm.

We notice in passing that $\Phi(1) \simeq 0.84$, so that the unshaded area in Fig. 10.10 is about 0.16 and hence $P(-1 \leqslant z \leqslant 1) \simeq 1 - 2(0.16) = 0.68$. But $P(-1 \leqslant z \leqslant 1) = P(-1 \leqslant (x - \mu)/\sigma \leqslant 1) = P(\mu - \sigma \leqslant x \leqslant \mu + \sigma)$ for any x which is $N(\mu, \sigma^2)$. This means that there is a probability of just over $\frac{2}{3}$ that any reading from a normal distribution lies within one standard deviation of its mean. Similarly using the value for $\Phi(2)$ which we found in the last example, $P(-2 \leqslant z \leqslant 2) = 1 - 2\Phi(2) \simeq 1 - 2(0.0228) \simeq 0.95$. So, again translating to the general case, we find that there is a probability of approximately 0.95 that any reading from a normal distribution lies within 2 standard deviations of the mean. Both these figures are useful ones to remember.

We can also use the table of $\Phi(z)$ the other way round; for instance, if the zoologist wished to know what length would be exceeded by 95% of the snakes, we would require x so that $F(x) = 0.05$ and hence $\Phi((x - 750)/150) = 0.05$. The table only gives positive values of z for which $\Phi(z) > 0.5$ but we can see that $\Phi(1.64) \simeq 0.95$ and hence
$$\Phi(-1.64) \simeq 0.05.$$
$$\therefore \frac{x - 750}{150} = -1.64 \qquad \therefore x = -246 + 750 = 504.$$

Hence approximately 95% of the snakes have lengths greater than 504 mm. (The value of 1.64 is confirmed from table 2 on page 244. The wording under the table explains why 1.64 corresponds to $P = 10$.)

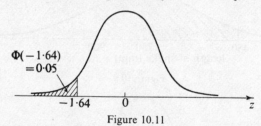

Figure 10.11

EXERCISE 10A

1. If x has the distribution $N(0, 1)$, find the probabilities of the following:
  (i) $x \leqslant 1.15$,
  (ii) $x \leqslant -0.65$,
  (iii) $0 \leqslant x \leqslant 1.15$,
  (iv) $-0.65 \leqslant x \leqslant 1.15$,
  (v) $|x| \leqslant 0.65$.

2. x is distributed normally with mean 3 and standard deviation 4. Find the probability that:
  (i) $x \leqslant 4$, (ii) $2 \leqslant x \leqslant 5$, (iii) $2x + 1 \leqslant 15$, (iv) $|x - 4| \geqslant 8$.
  Show that 95% of the distribution lies between -4.84 and 10.84.

3. Find $P(x \geqslant 0)$ when x has the distribution (i) $N(-2, 25)$, (ii) $N(2, 25)$.

4. Find the probability that a normally distributed variable differs from its mean by more than three times its standard deviation.

5. Use Simpson's rule or another numerical method of integration to show that
$$\frac{1}{\sqrt{(2\pi)}} \int_{-\infty}^{\infty} e^{-x^2/2} dx = 1.$$

(In practice you will find that it is sufficient to consider the region from $x = 0$ to $x = 4$. The use of a calculating machine or computer will help.)

6. A machine produces components to any required length specification with a standard deviation of 1·40 mm. At a certain setting it produces to a mean length of 102·30 mm. Assuming the distribution of lengths to be normal calculate:
  (i) what percentage would be rejected as less than 100 mm long,
  (ii) to what value, to the nearest 0·01 mm, the mean should be adjusted if this rejection rate is to be 1 %,
  (iii) whether at the new setting more than 1 % of components would exceed 107 mm in length. (M.E.I.)

7. To what value in the distribution $N(50, 100)$ would a value of 50 in $N(55, 225)$ correspond? Give a formula for converting marks in an examination distributed as $N(55, 225)$ to one with $N(50, 100)$.

8. Goodlite state that their Longlife Photoflood lamps have an average life of 3 hours and that 97·7% of the bulbs have a life of at least 2 h. Estimate the standard deviation. What percentage of the bulbs would you expect:
  (a) to last more than $3\frac{1}{2}$ h, (b) to fail in less than $2\frac{1}{4}$ h?
What assumptions have you made about the distribution? (Cam.)

9. Show that the points of inflection of the curve
$$y = \frac{1}{\sqrt{(2\pi)}\sigma} e^{-\frac{1}{2}(x-\mu)^2/\sigma^2}$$
occur at $x = \mu \pm \sigma$.

10. Observation of a very large number of cars at a certain point on a motorway establishes that the speeds are normally distributed. 90% of cars have speeds less than 124·3 km/h, and only 5% of cars have speeds less than 101·0 km/h. Determine the mean speed μ and the standard deviation σ. (Cam.)

11. Ball bearings are manufactured with a nominal diameter of 2 mm, but are acceptable if their diameters are inside the limits 1·90 mm to 2·10 mm. It is observed that, in a large batch, $2\frac{1}{2}$% are rejected as oversize and $2\frac{1}{2}$% as undersize. Assuming that the diameters are normally distributed, approximately what proportions will be rejected if the limits are changed to 1·95 mm and 2·15 mm? (J.M.B.)

12. In the manufacture of machined components the proportion of the product which does not pass through a gauge of 4·025 cm is 6% and the proportion passing a gauge of 3·965 cm is 2%. What proportion of the product lies in the range $4·000 \pm 0·0275$ cm if the dimension of the component is normally distributed? (M.E.I.)

13. If x is distributed as $N(\mu, \sigma^2)$, prove that
$$P(ax+b \leqslant z) = \Phi\left(\frac{z-a\mu-b}{a\sigma}\right)$$
and hence deduce the distribution of $ax+b$.

14. The diameters of some machined components are distributed normally with mean 5·00 units and standard deviation 0·05 units. Find the expected proportion of the components which will be outside the range 4·925 units to 5·075 units and the ratio of the expected proportion in the range $5·025 \leqslant d \leqslant 5·050$ to the expected proportion in the range $5·050 \leqslant d \leqslant 5·075$. It is desired to adjust the mean of the process so that there are, on average, twice as many components in the range 5·025 units to 5·050 units as in the range 5·050 units to 5·075 units. Show that this can be done if the mean is adjusted to a value between 5·00 units and 4·95 units, and find the value by trial to the nearest 0·01 unit. (M.E.I.)

15. A man who was trying out various makes of razor blade kept a record of the number of shaves x he got from each blade before he had to discard it. For 50 blades of one make he found that x was approximately normally distributed with mean 7·2 and standard deviation 2·0. Draw the frequency curve for this distribution. For 50 blades of a different make he obtained the following distribution

x	5	6	7	8	9	10	11	12	13	14
Frequency	1	2	4	6	9	9	8	6	3	2

On the same diagram as before draw a histogram of the same area to represent the new distribution. (You should draw the histogram by treating $x = 5$ as $4·5 \leqslant x \leqslant 5·5$, etc.) Describe in words any inference that can be made from the two distributions.

(J.M.B.)

16. An investigation into the weekly spending money of the 500 boys in a certain school showed that the mean was 34p and the standard deviation 5p. Assuming that the distribution was normal, calculate the theoretical frequencies for the intervals 20·5p to 25·5p, 25·5p to 30·5p . . . 45·5p to 50·5p. Draw the histogram.

17. Certain components are manufactured to have a mean diameter of 10 cm and standard deviation 0·1 cm. Find values a and b so that 95% of the components have diameters within $10 \pm a$ cm, and 99% of the components have diameters within $10 \pm b$ cm. The manufacturing process is said to be "out of control" if, when a sample of components are selected, at random, from the production line, either one has diameter outside $10 \pm b$ cm or two consecutively have diameters outside $10 \pm a$ cm. If a particular sample had diameters 10·04, 10·15, 9·90, 10·16, 10·17, 10·03, 9·80, 9·78, 10·10, 10·15, determine whether or not the process was "out of control". Illustrate your answer by plotting a graph with diameter measured along the y-axis and the sample points equally spaced along the x-axis and marking in the lines $y = 10 \pm a$ and $y = 10 \pm b$. (Such a graph is known as a *control chart*.)

18. The weights of Granny Smith apples produced by a grower are normally distributed with mean 110 g and standard deviation 10 g. Only those apples above average weight can be sold as first-grade apples. What is the average weight of a first-grade apple?

19. The marks in an examination are $N(50, 100)$. It is desired to scale the marks so that the pass-mark is 40 and 70% of the candidates pass, the distinction mark is 70 and 20% of the candidates get distinction. Calculate the new mark of a candidate who had an original mark of 60.

20. Engine crankshafts are manufactured so that the diameters, in centimetres, form a normal distribution with mean 5 and standard deviation 0·03. Crankshafts with diameters less than 4·94 or greater than 5·06 are rejected. The accepted product is classified into three grades of size, 4·940 to 4·988, 4·988 to 5·012, 5·012 to 5·060. Show that:

(i) $\int xe^{-\frac{1}{2}x^2}dx = -e^{-\frac{1}{2}x^2}$,

(ii) $\int x^2e^{-\frac{1}{2}x^2}dx = -xe^{-\frac{1}{2}x^2} + \int e^{-\frac{1}{2}x^2}dx$.

Hence find the average diameter in each of the three grades, and the ratio of the standard deviation in the middle grade to the standard deviation of the unclassified product before any rejection of under- and over-size shafts.

(M.E.I.)

10.5. The normal approximation to the binomial distribution

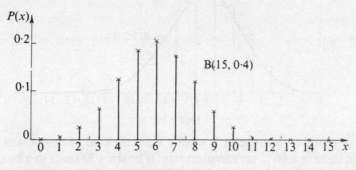

Figure 10.12

Figure 10.12 shows the probability function $B(15, 0\cdot4)$. If we draw some similar probability graphs for the binomial distribution for values of p close to $\frac{1}{2}$, we will notice that the distribution is almost symmetric and has its largest probabilities for central values of the variable x. If in addition n is large, so that x can take many different values, the graphs have the characteristic shape of the normal distribution even though x is discrete.

The similarity between the graphs of the binomial and the normal distributions is even more marked if the binomial graphs are drawn as histograms with the probability for $x = 3$, say, spread out evenly over the range $2\cdot5$–$3\cdot5$.

It does appear, and it can be theoretically proved, that for p close to $\frac{1}{2}$ and n large we can approximate to $B(n, p)$ by an appropriate normal distribution. Clearly this should be the normal distribution with $\mu = np$ and $\sigma^2 = npq$, since these are the actual mean and variance of the binomial distribution, and so we should approximate to $B(n, p)$ by $N(np, npq)$; for instance, we would approximate to $B(15, 0\cdot4)$ by $N(6, 3\cdot6)$. However, since we are now treating a discrete variable as if it were continuous, we should do it in the way suggested by the histograms, which means that we should regard $x = 3$ for the binomial as corresponding to $2\cdot5 \leqslant x \leqslant 3\cdot5$ for the normal approximation. This will imply that if we want to find a probability of the form $P(x \leqslant 3)$ we should approximate to this by $P(y \leqslant 3\cdot5)$ (where y has the normal distribution) and to $P(x \geqslant 3)$ by $P(y \geqslant 2\cdot5)$. Figure 10.13 shows $B(15, 0\cdot4)$ plotted as a histogram with the corresponding $N(6, 3\cdot6)$ superimposed on it.

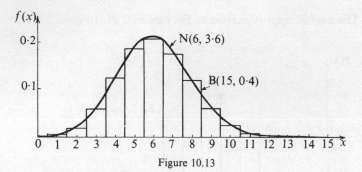

Figure 10.13

For example, a gardener is sowing a certain kind of flower seed which is said to have a 60 % germination rate. If he sows 30 seeds in a box, what is the probability that 15 or less will germinate? The number of seeds which germinate, x, will have the distribution $B(30, 0.6)$ and we require $P(x \leqslant 15)$ which would be given by

$$\sum_{x=0}^{15} \binom{30}{x}(0.6)^x(0.4)^{30-x}.$$

To calculate this as it stands would be a formidable task. However, since $p = 0.6$ which is fairly close to $\frac{1}{2}$ and $n = 30$ which is quite large, we can use the approximation of the normal distribution with $\mu = 30 \times 0.6 = 18$ and variance $30 \times 0.6 \times 0.4 = 7.2$, i.e. $N(18, 7.2)$.

So we can say $P(x \leqslant 15) \simeq \Phi\left(\dfrac{15.5 - 18}{\sqrt{7.2}}\right) \simeq \Phi(-0.93) = 1 - \Phi(0.93)$

$\simeq 1 - 0.824 = 0.176.$

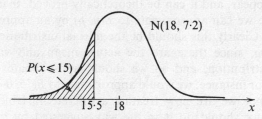

No. of seeds which germinate out of 30

Figure 10.14

Hence the chance that half of the seeds or less will germinate is about one sixth. We can see that the calculation is very much quicker than the one we would have had to do for the actual binomial distribution.

What values should n and p have before we can use the approximation safely? We have so far just said that it is valid if n is "large" and p is "close" to $\frac{1}{2}$. If p is near $\frac{1}{2}$, say $0.4 < p < 0.6$, the approximation is reasonably good for n greater than 10. In general the further p is from $\frac{1}{2}$ the larger n needs to be. However if n is very large, the approximation is quite accurate even for values of p close to 0 or 1.

We have now met two approximations to the binomial distribution: the Poisson distribution for small p and the normal for more general p. The normal is the more fundamental of the two approximations and is a special case of a general result which we shall meet in the next chapter. It can indeed be proved that, provided μ is large, the Poisson distribution $\mathscr{P}(\mu)$ can also be approximated by a normal distribution. The appropriate one to use will be $N(\mu, \mu)$ since both the mean and variance of $\mathscr{P}(\mu)$ are equal to μ.

EXERCISE 10B

1. Twenty pennies are tossed together a large number of times. State the theoretical mean and the standard deviation of the distribution of the number of heads. Assuming that the distribution closely approximates to the normal curve, use tables to find the probability that 14 or more heads will appear together in a single toss. (J.M.B.)

2. The police force in a certain district carries out tests on the brakes of automobiles chosen at random on the road. Each man is required to test 20 cars. Calculate the distribution of the number of cars with defective brakes in sets of 20 cars if the probability that a single car has defective brakes is 10%. Show the distribution graphically together with a plot of the normal distribution with the same mean and variance. Comment on the relation between the two distributions. (M.E.I.)

3. A machine produces articles with an average of 20% which are defective. Find an approximate value for the probability that a sample of 400 items will contain more than 96 which are defective. (J.M.B.)

4. A die is thrown 300 times, a score of 1 or 2 being counted as a "success". Use the normal approximation to the binomial distribution to derive the probability that the number of "successes" will not deviate from 100 by more than 15.
(J.M.B.)

5. In a certain large town Labour usually polls $\frac{1}{2}$ of the votes cast. Each of 20 investigators asks a random sample of 16 voters which way they will vote. Calculate the standard deviation for such a sample of 16 voters. Use the corresponding normal curve approximation to calculate the number of investigators whom you would expect to report that less than 6 of their sample say that they will vote Labour. (J.M.B.)

6. The number of serious road accidents in a given town in a given month can be assumed to have the Poisson distribution with mean 20. Calculate approximately the probability that last month there were:
   (i) less than 25 accidents,
   (ii) 18 or more accidents.

7. A certain component in a continuously running machine fails and has to be replaced at an average rate of once every 24 hours. Assuming that the total number of failures in any fixed time interval follows a Poisson distribution, with a mean value proportional to the length of the time interval, find the number of spare components which should be stocked so that the risk that the stock will be exhausted during any particular week will not exceed 5%. (You may use the normal approximation.) (J.M.B.)

REVISION EXAMPLES III

1. Two players compete by drawing in turn and without replacement one ball at random from a box containing four red and four white balls. The winner is the player who first draws a red ball. Calculate the probability that the winner is the player who makes the first draw. Derive the corresponding probability when the balls are replaced after each draw. For the case when the balls are not replaced, calculate the mean number of balls that will be drawn in such a competition. (J.M.B.)

2. The life-times, x hours, of radio valves have the continuous probability density function $f(x) = \frac{1}{1000}e^{-x/1000}$, $x \geqslant 0$. An amplifier contains five valves, the life-time of each having the above distribution. The makers guarantee that not more than two valves will have to be replaced during the first 1000 hours of use. Find the probability that the guarantee is violated, assuming that the valves wear out independently. (J.M.B.)

3. A man leaves home at 8 am every morning in order to arrive at work at 9 am. He finds that over a long period he is late once in forty times. He then tries leaving home at 7.55 am and finds that over a similar period he is late once in one hundred times. Assuming that the time of his journey has a normal distribution, at what time should he leave home in order not to be late more than once in two hundred times? (S.M.P.)

4. A probability distribution has the probability density function $f(x) = ke^{-\lambda x}$ $x \geqslant 2$, k constant. Find the mean, median and standard deviation of the distribution. (M.E.I.)

5. A mass-produced circular disc should have radius 2 cm, but in fact the values of the radius are uniformly distributed in the range 1·95–2·10 cm. Explain why
(a) median of the area = π (median of the radius)^2,
(b) mean of the area $\neq \pi$ (mean of the radius)^2.
Obtain limits A_L and A_U for the area such that 20% of the discs have area less than A_L and 20% have area greater than A_U. (Cam.)

6. In a simplified probability model of the service in a barber's shop it is supposed that all haircuts take exactly six minutes and that a fresh batch of customers arrives at six-minute intervals. The number of customers in a batch is described by a Poisson probability function, the mean number being 3. Any customer who cannot be served instantly goes away and has his hair cut elsewhere. The shop is open for forty hours a week. Calculate the theoretical frequencies with which batches of 0, 1, 2, 3, 4, 5 and more than 5 customers will arrive. The proprietor reckons that is costs him £25 a week to staff and maintain each chair in his shop, and he charges 25p for each hair-cut. Calculate his expected weekly profit if he has: (i) 3, (ii) 4, (iii) 5 chairs. (S.M.P.)

7. Through a fixed point P on a circle of unit radius a random chord is drawn, such that the angle between the chord and the diameter through P is uniformly distributed in $(-\frac{1}{2}\pi, \frac{1}{2}\pi)$. Show that the distribution of the length of the larger of the two parts into which the circumference is divided is uniform in $(\pi, 2\pi)$. Find also the mean value of the area of the larger of the two parts of the circle, and the mean length of the chord. (J.M.B.)

8. In a particular school 1460 pupils were present on a particular day. By 8.40 am 80 pupils had already arrived, and at 9.00 am 12 pupils had not arrived but were on their way to school. By assuming that the frequency function of arrival times approximates to normal form, use tables to estimate:
   (i) the time by which half of those eventually present had arrived,
   (ii) the standard deviation of the times of arrival.
   If registration occurred at 8.55 am, how many would not have arrived by then? If each school entrance permitted a maximum of 30 pupils per minute to enter, find the minimum number of entrances required to cope with the "peak" minute of arrival. (S.M.P.)

9. A random variable X has the probability density function

$$f(x) = \frac{\pi e^{-\pi x/\sigma\sqrt{3}}}{\sigma\sqrt{3}\{1 + e^{-\pi x/\sigma\sqrt{3}}\}^2}$$

   for all real x. Show that the distribution is symmetrical about $x = 0$. Determine the cumulative distribution $F(x) = P(X \leqslant x)$. You are given that the variance of X is σ^2. Compare the values of $F(\sigma)$, $F(2\sigma)$, $F(3\sigma)$ with the corresponding values for the normal distribution with zero mean and variance σ^2.
(M.E.I.)

10. If a currant bun is chosen at random in a baker's shop, the probability p_r that it contains r currants is given by $p_0 = 0.02, p_1 = 0.08, p_2 = 0.16, p_3 = p_4 = 0.22$. A shopper buys six buns which are chosen at random from the buns in the shop.
   (i) Calculate the probability that none of the buns contains more than two currants.
   (ii) Calculate the mean and variance of the number of buns with five or more currants.
   (iii) Calculate the probability that exactly two buns contain five or more currants. (Cam.)

11. The traffic lights at a road obstruction work on a cycle of $\frac{1}{2}$ min at GO followed by $1\frac{1}{2}$ min at STOP. Find the probability that a car arriving at a random instant of time will be stopped by the lights. If the random variable T is the length of time in minutes which the car has to wait in order to pass the lights, find the cumulative distribution function and the mean and variance of T. (Cam.)

12. From a batch of manufactured articles a sample of ten is taken and each article is examined. If two or more articles are found to be defective, the batch is rejected; otherwise it is accepted. Show that, if p is the proportion defective in a batch and P its chance of being accepted, $P = (1-p)^9(1+9p)$. Find an expression for P if it is now decided to modify the scheme so that, when one defective is found in the sample, a second sample of ten is taken and the batch rejected if this sample contains any defectives. In the second case what will be the average number sampled per batch over a large number of batches when $p = 0.05$? (M.E.I.)

13. A random variable has density function of the form $f(x) = 0$ $(x < 0)$, $f(x) = e^{-ax}(1 - e^{-bx})$ $(x \geqslant 0)$ where a and b are constants and $a > 0$. Show that $b > 0$, and express b in terms of a. Deduce that $0 < a < 1$. For the case $a = b$, sketch the density function and calculate the mean and variance of the random variable.

           (J.M.B.)

14. In the equation $x^2 + 2x - a = 0$, a has a rectangular distribution on the interval $(0, 2)$. Find the distribution of the larger root. Find the probability that, of six of these equations in each of which a is taken independently from the rectangular distribution on $(0, 2)$, no more than one has its larger root greater than $\sqrt{2} - 1$.

           (J.M.B.)

15. A grocer sells bread and can buy batches of 120 loaves. The number of daily customers for bread is distributed normally with mean 100 and variance 100. The net profit on the sale of a loaf is 1p, and the net loss on an unsold loaf is 1·5p. What is the average daily net profit to the grocer?

           (M.E.I.)

16. The diagram shows the graph of a probability density function with mean 5. Suggest a suitable equation for it involving an exponential function, and calculate the standard deviation.

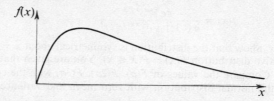

           (S.M.P.)

17. The moment generating function (m.g.f.) of a variable x is defined as $M(t) = E(e^{xt})$, where $E(.)$ is the expectation operator. If $M(t)$ is known in the form of an expansion in ascending powers of t, show how the mean and variance of x can be determined from it. Find the m.g.f. of x when the frequency function is $f(x) = \frac{1}{2}$, $1 \leqslant x \leqslant 3$. Hence find the variance of x and verify your result from the definition of variance.

           (Cam.)

18. The 8 football clubs who survived to the fifth round of the F.A. cup this year were 4 First Division teams and 4 Second Division teams. The draw for the round is made by placing the names in a hat and drawing them out one by one. The first team drawn is at home to the second team drawn and so on. Find:

  (i) the probability function of the number of Second Division teams to play at home and verify that the expected number is 2,

  (ii) the expected number of matches featuring one First Division and one Second Division team.

What would be the answers to (i) and (ii) if there were 5 First Division and 3 Second Division teams?

19. I plant 6 daffodils in a tub. The daffodils grow and flower independently. The probability that a daffodil grows at all is p. If it does grow, the number of flowers it has is distributed as $\mathscr{P}(\mu)$. Find:

  (i) the probability that I have no flowers in my tub,

  (ii) the expected number of daffodils that have at least one flower,

  (iii) the expected number of flowers on each daffodil.

Show that the probability that I get 3 flowers in my tub is

$$pe^{-\mu}\mu^3(pe^{-\mu} + q)^3(36p^2e^{-2\mu} + 17pqe^{-\mu} + q^2), \text{ where } q = 1 - p.$$

20. During World Cup year a certain petrol company gave away a set of 30 medals bearing the portraits of members of the English team. They gave one medal for each four gallons of petrol purchased, the medals being chosen at random. If I bought 12 gallons of petrol each week, find the expected number of different medals I received the first week. Given that I had already collected n different coins one week, find the expected number of new and different medals I received the next week.

Estimating from Samples

11.1. The binomial distribution

In the last four chapters we have been discussing probability models which might apply to experimental situations. We now want to decide how to set about fitting any particular model to a given set of data. In later chapters we will find out how to tell whether the model really fits and, if it does, what conclusions can be drawn from a limited amount of data.

Fitting a model to a situation on the basis of sample data takes two steps, of which the first is the selection of the type of distribution to be fitted. We have to ask such questions as: "Are the data discrete or continuous?" or "Is the situation a success/failure type?" and plot a histogram to see the general shape of the distribution. The second step is to assign values to any unknown parameters in the chosen distribution. When choosing a distribution there are few hard-and-fast rules, and often several different distributions have to be tried before a best-fitting one is found.

As an example, suppose we take one of the problems we met in the first chapter, of how to find the probability that a man chosen at random would vote Labour if there were an election next week. We decided there that we would have to base our figure on a sample, as it would clearly be impracticable to question the whole population. Suppose we took a sample of 100 men and found that 45 stated an intention to vote Labour; it then seems natural to take 0·45 or 45 % as our estimate of the probability that a randomly chosen male voter will vote Labour. We could simply rest content with this estimate, but it would be of little use unless we knew how accurate it was likely to be. In order to do this we need to imagine what would happen if we took a large number of different samples, each of 100 voters, and found the number of Labour supporters in each, which we shall call r. We would therefore obtain a distribution of the random variable r over the large number of samples. This distribution is known as the **sampling distribution** of r, and r itself is technically called a **statistic**, as it is a quantity whose value depends only on the data from a sample.

If we assume that the sample is a random one, then we have a success/failure situation with 100 trials and thus we would expect r to have at any rate approximately the sampling distribution $B(100, p)$, where p is the probability of a random man voting Labour. Notice that as p is the probability that we are trying to estimate, we are in fact in a circular situation where the distribution of our estimate depends on the unknown quantity itself.

We are now in a position to say what would happen if we took a large number of random samples of 100 and each time worked out our estimate of p, which we shall call $\hat{p}$. It seems reasonable that we would use $r/100$ as our value of $\hat{p}$ for any one sample. (You will often find $\hat{p}$ called an *estimator* and its value for a particular sample an *estimate*. We will generally call both estimates.)

Hence $E(\hat{p}) = E(r/100) = \dfrac{E(r)}{100} = \dfrac{100p}{100} = p$, as r is $B(100, p)$.

(Notice that here we have used $E(r/100) = \frac{1}{100}E(r)$ which is a special case of the result $E(ax) = aE(x)$; this is easy to prove if you have not already done so. Since the variance is the expectation of a squared quantity, the corresponding result will be $\text{Var}(ax) = a^2\text{Var}(x)$.)

So we can at least see that the expected value of our estimate is p; in other words on average our estimate is giving us the right result. We therefore say that $\hat{p}$ is an *unbiased* estimate of p.

How close is our estimate likely to be to p, which we have just seen is its mean? We might measure this by $E[(\hat{p}-p)^2]$ which is $\text{Var}(\hat{p})$ as $\hat{p}$ is unbiased. We would hope that, if the estimate is a good one, it would have a small variance, so that we are unlikely to obtain an estimate which is far from p.

$$\text{Var}(r/100) = \frac{\text{Var}(r)}{100^2} = \frac{100pq}{100^2} = \frac{pq}{100}$$

The variance is therefore less than $\frac{1}{100}$ since we know p and q are less than 1. In the general case for a sample of size n we have

$$\hat{p} = r/n \qquad E(\hat{p}) = E(r/n) = \frac{np}{n} = p \text{ (again the estimate is unbiased)}.$$

$$E[(\hat{p}-p)]^2 = \text{Var}(\hat{p}) = \text{Var}(r/n) = \frac{npq}{n^2} = \frac{pq}{n}.$$

So we can see that the variance decreases, and therefore the accuracy of the estimate increases, as the sample size increases. Furthermore $E[(\hat{p}-p)^2]\to 0$ as $n\to\infty$. Such an estimate is said to be *consistent*, and it is clearly desirable that any estimate we use should be both unbiased and consistent.

There are many other properties of estimates which we could have discussed, but these two are among the most important. In general we shall choose estimates which are both unbiased and consistent. If we have a case with two or more possible estimates with these properties, we will prefer the one with the smallest variance.

Now let us return to our actual example; we have shown that if we take 100 voters and find that r of them intend to vote Labour, then $r/100$ is an unbiased and consistent estimate of the probability that any voter will vote Labour. We also note that r would be distributed as $B(100, p)$ if we were to take many samples each of size 100, and that our estimate would have variance pq/n. The best estimate we can make of this variance would be $\hat{p}\hat{q}/n = \hat{p}(1-\hat{p})/n$.

So in the particular case with $r = 45$ we have $\hat{p} = 0\cdot45$ and the best estimate we can make of its variance is $\dfrac{0\cdot45\times0\cdot55}{100} \simeq \frac{1}{400}$, giving a standard deviation for $\hat{p}$ of roughly $\frac{1}{20}$ or $0\cdot05$. We can see that if the estimate was based on a sample of 10 000 voters, of whom 4500 intend to vote Labour, the standard deviation of the estimate would be estimated at $\frac{1}{200}$ or $0\cdot005$.

11.2. Estimation of mean and variance

The managing director of a company about to launch a new product decides to commission a number of forecasts of the size of the possible market, which we shall call $x_1, x_2, \ldots x_n$. What would be his best estimate of the market, based only on this set of forecasts, and what margin of error is there likely to be. between this estimate and the real figure?

Again in order to make any progress we will have to make certain assumptions. It may be that all the forecasts for some reason greatly overestimate the potential market, in which case there would be few reliable conclusions we could draw from them. However, if we assume that the forecasts are liable to random rather than systematic error, and are made by people equally competent to judge the situation, it would seem reasonable to accept that each forecast might come from a distribution with mean μ, where μ is the actual market size (which is unknown) and variance σ^2, where σ^2 is the variance of the total population of such forecasts of the market size, and is hence also unknown.

This means that if x_i is a typical forecast, we are assuming that $E(x_i) = \mu$ and $\text{Var}(x_i) = \sigma^2$. We will also assume that the forecasts are independent of each other, thus they are a *random sample* of all forecasts.

What we want to obtain from the forecasts is an estimate of μ. There is no obvious reason why we should not take the median of the sample of forecasts as our estimate of μ (or for that matter the mode if there is one), but it is a fair guess that most people would choose to use the sample mean $\bar{x}$. If we also wanted an estimate of σ^2, we might well choose to use the sample variance

$$S^2 = \sum_i \frac{(x_i - \bar{x})^2}{n}$$

Are $\bar{x}$ and S^2 good estimates of μ and σ^2 in the sense of being un-biased and consistent? The answer is "yes" for one estimate and "no" for the other, as we shall see shortly.

The four questions we need to answer are:
Does $E(\bar{x}) = \mu$? If so, $\bar{x}$ is an unbiased estimate of μ.
Does $\text{Var}(\bar{x}) \to 0$ as $n \to \infty$? If so, $\bar{x}$ is also a consistent estimate of μ.
Does $E(S^2) = \sigma^2$?
Does $\text{Var}(S^2) \to 0$ as $n \to \infty$?

11.3. The expectation and variance of a sum of random variables

To investigate the first two queries we need

$$E(\bar{x}) = E\left[\frac{(x_1 + x_2 + \ldots + x_n)}{n}\right] = \frac{1}{n}E(x_1 + x_2 + \ldots + x_n)$$

and $\text{Var}(\bar{x}) = \text{Var}\left[\frac{(x_1 + x_2 + \ldots + x_n)}{n}\right] = \frac{1}{n^2}\text{Var}(x_1 + x_2 + \ldots + x_n).$

Hence we need to find $E(x_1 + x_2 + \ldots + x_n)$ and $\text{Var}(x_1 + x_2 \ldots + x_n)$ where we know that each variable x has mean μ and variance σ^2.

Suppose we look first at the simpler case, which arose in Exercise 7A, of throwing 2 dice and calculating the sum x of the two scores. We should have found that the probability distribution was:

x	2	3	4	5	6	7	8	9	10	11	12
$P(x)$	$\frac{1}{36}$	$\frac{2}{36}$	$\frac{3}{36}$	$\frac{4}{36}$	$\frac{5}{36}$	$\frac{6}{36}$	$\frac{5}{36}$	$\frac{4}{36}$	$\frac{3}{36}$	$\frac{2}{36}$	$\frac{1}{36}$

and hence calculated $E(x) = 7$ and $\text{Var}(x) = \frac{35}{6}$.

We could, however, write x as $x_1 + x_2$ where x_1 is the score on the first die and x_2 is the score on the second die. Notice that the value of x_1 is completely unaffected by the value of x_2 and vice versa; so the variables are *independent*.

We know that both x_1 and x_2 have the same uniform distribution $P(x) = \frac{1}{6}$, $x = 1 \ldots, 6$ and can therefore calculate

$$E(x_1) = E(x_2) = 3\cdot5 \text{ and } \mathrm{Var}(x_1) = \mathrm{Var}(x_2) = \tfrac{35}{12}.$$

Thus it appears that, at least in this example,

$$E(x) = E(x_1 + x_2) = E(x_1) + E(x_2) = 2E(x_1)$$

and $\qquad \mathrm{Var}(x) = \mathrm{Var}(x_1 + x_2) = \mathrm{Var}(x_1) + \mathrm{Var}(x_2) = 2\,\mathrm{Var}(x_1).$

Similarly we could work out that if we threw n dice and found the sum of their scores x, we would have

$$\begin{aligned}
E(x) &= E(x_1 + x_2 + \ldots + x_n) \\
&= E(x_1) + E(x_2) + \ldots + E(x_n) = nE(x_1) = 7n/2. \\
\mathrm{Var}(x) &= \mathrm{Var}(x_1 + x_2 + \ldots + x_n) \\
&= \mathrm{Var}(x_1) + \mathrm{Var}(x_2) + \ldots + \mathrm{Var}(x_n) \\
&= n\mathrm{Var}(x_1) = 35n/12.
\end{aligned}$$

All we have given here is one simple illustration of a mathematical result that can be proved theoretically, which is that

$$\begin{aligned}
E(x_1 + x_2 + \ldots + x_n) &= E(x_1) + E(x_2) + \ldots + E(x_n) \\
&= nE(x_1)
\end{aligned}$$

(if all the values of x have the same mean).

$\mathrm{Var}(x_1 + x_2 + \ldots + x_n) = \mathrm{Var}(x_1) + \mathrm{Var}(x_2) + \ldots + \mathrm{Var}(x_n)$, if all the variables x are *independent* (i.e. none of them is affected by any of the others) $\qquad\qquad = n\,\mathrm{Var}(x_1)$

(if in addition they all have the same variance).

If we now return to our estimate $\bar{x}$, we have

$$E(\bar{x}) = \frac{1}{n} E(x_1 + x_2 + \ldots + x_n) = \frac{1}{n}\{E(x_1) + E(x_2) \ldots + E(x_n)\} = \frac{n\mu}{n} = \mu.$$

Hence $\bar{x}$ is an unbiased estimate of μ.

Also $\mathrm{Var}(\bar{x}) = \dfrac{1}{n^2}\,\mathrm{Var}(x_1 + x_2 + \ldots + x_n)$

$$= \frac{1}{n^2}\{\mathrm{Var}(x_1) + \mathrm{Var}(x_2) + \ldots + \mathrm{Var}(x_n)\} = \frac{n\sigma^2}{n^2} = \frac{\sigma^2}{n}$$

Since $\mathrm{Var}(\bar{x}) \to 0$ as $n \to \infty$, $\bar{x}$ is also a consistent estimate of μ.

We can thus use the mean of a random sample of measurements as an estimate of the mean of the population with the assurance that this estimate is both unbiased and consistent.

11.4. The standard error

Let us look again at the last result $\mathrm{Var}(\bar{x}) = \sigma^2/n$, which holds for any random sample n of independent measurements from a population with variance σ^2. (Strictly to ensure independence the population should be infinite or *sampled with replacement*, but if the population is a large one the result will be approximately true even for *sampling without replacement*.) This tells us that the larger the sample size, the smaller the variance of our estimate and therefore the greater its accuracy. For instance the variance of the means of many samples of size 50 is likely to be about $\frac{1}{5}$ of that of samples of size 10 from the same distribution. Notice that the standard deviation of the mean of a sample of size n is inversely proportional to the *square root* of the sample size as

$$\text{standard deviation of } \bar{x} = \frac{\sigma}{\sqrt{n}}.$$

This quantity $\sigma/\sqrt{n}$ is often referred to as the **standard error of the mean** as it gives a measure of the accuracy of the sample mean as an estimate of the population mean.

11.5. The distribution of $\bar{x}$: the central limit theorem

We have now found that the means of random samples of size n from a population with mean μ and variance σ^2 will have a distribution with mean μ and variance σ^2/n. But what sort of a distribution will it be? In general we would expect the distribution to depend on the original distribution of the population.

The answer is particularly simple if the population is $N(\mu, \sigma^2)$ for then we can say that $\bar{x}$ will also be normally distributed in the $N(\mu, \sigma^2/n)$ distribution. This is an important result and one which we shall often refer to in future (see Section 11.7); we shall not, however, give the proof as it involves the probability density function of the normal distribution which, as we have seen, is difficult to handle mathematically.

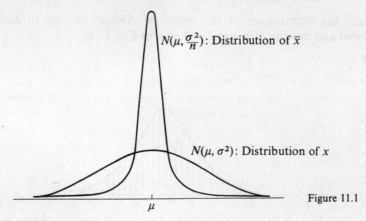

$N(\mu, \frac{\sigma^2}{n})$: Distribution of $\bar{x}$

$N(\mu, \sigma^2)$: Distribution of x

μ

Figure 11.1

What if the population had a different distribution? Then it is difficult in general to derive an exact result, but we can arrive at an approximate one for large values of n.

Suppose we return to the case where x is the score on a die and is uniformly distributed.

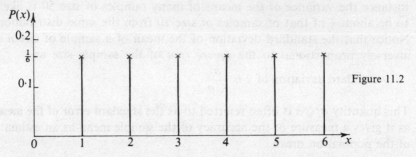

Figure 11.2

What happens when we take the mean score of a sample of size 2? This is simply half the sum of the scores on the two dice which has the distribution shown in Fig. 11.3.

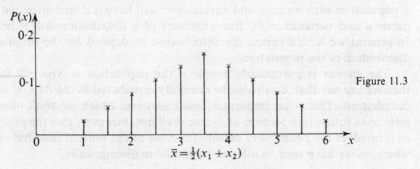

Figure 11.3

$$\bar{x} = \tfrac{1}{2}(x_1 + x_2)$$

Similarly the distribution of the mean of samples of size 10 can be calculated and has the distribution shown in Fig. 11.4.

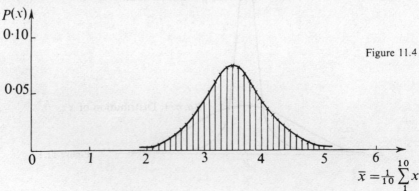

Figure 11.4

$$\bar{x} = \tfrac{1}{10}\sum_1^{10} x_i$$

(It is worth trying this experimentally for different sample sizes by collecting results from a very large number of throws and collecting them into samples of different sizes.)

If we looked at results for other sample sizes as well, we could see that the larger the sample size the nearer the distribution gets to the characteristic shape of the normal distribution, even when the original values come from a uniform distribution.

This example illustrates one of the most fundamental statistical laws known as the **central limit theorem,** which states that the distribution of the mean $\bar{x}$ of random samples of size n from any population with mean μ and finite variance σ^2 tends to $N(\mu, \sigma^2/n)$ as $n \to \infty$.

In practical terms we can usually assume that the distribution of $\bar{x}$ is approximately $N(\mu, \sigma^2/n)$ for $n > 20$, for any distribution of x which is not too skew. The approximation is best when x is nearly normally distributed and can then be used for smaller n.

11.6. The expectation of S^2

Suppose we now return to our original problem of estimating the potential market from a set of forecasts. We have seen that the mean $\bar{x}$ gives an unbiased and consistent estimate of the actual market, but we do not know how accurate an estimate it is. We know that the standard error of $\bar{x}$ is $\sigma/\sqrt{n}$ and hence we need to be able to estimate σ before we can give a rough figure for the likely accuracy.

We suggested S^2 as a possible estimate for σ^2.

What is $E(S^2)$? We shall show that

$$E(S^2) = \frac{(n-1)\sigma^2}{n}, \text{ though the proof is complicated.}$$

We shall find it easier to work with $\sum_i (x_i - \mu)^2 = \sum_i (x_i - \bar{x} + \bar{x} - \mu)^2$

$$= \sum_i (x_i - \bar{x})^2 + 2\sum_i (x_i - \bar{x})(\bar{x} - \mu) + \sum_i (\bar{x} - \mu)^2$$

When we do the summation, the quantity $(\bar{x} - \mu)$ is a constant so we have

$$\sum_i (x_i - \mu)^2 = \sum_i (x_i - \bar{x})^2 + 2(\bar{x} - \mu) \sum_i (x_i - \bar{x}) + n(\bar{x} - \mu)^2$$

But, from the definition of $\bar{x}$, $\sum_i (x_i - \bar{x}) = 0$

so we can write $\sum_i (x_i - \mu)^2 = \sum_i (x_i - \bar{x})^2 + n(\bar{x} - \mu)^2$

thus $\sum_i (x_i - \bar{x})^2 = \sum_i (x_i - \mu)^2 - n(\bar{x} - \mu)^2$,

i.e. $S^2 = \dfrac{\sum_i (x_i - \mu)^2}{n} - (\bar{x} - \mu)^2$ and so we have

$$E(S^2) = E\left[\frac{\sum_i (x_i - \mu)^2}{n} - (\bar{x} - \mu)^2\right]$$

By the results of section 11.3 we can write this as

$$E(S^2) = \frac{1}{n}\sum_i E[(x_i-\mu)^2] - E[(\bar{x}-\mu)^2]$$

Since we know each x_i has mean μ and variance σ^2, $E[(x_i-\mu)^2] = \text{Var}(x_i)$ $= \sigma^2$ and further, since $\bar{x}$ also has mean μ, $E[(\bar{x}-\mu)^2] = \text{Var}(\bar{x}) = \sigma^2/n$.

So $E(S^2) = \dfrac{n\sigma^2}{n} - \dfrac{\sigma^2}{n} = \dfrac{(n-1)\sigma^2}{n}$.

We can see that S^2 is not an unbiased estimate of σ^2. We can remedy this if, instead of using S^2, we take a new estimate

$$s^2 = \frac{nS^2}{n-1} = \frac{\Sigma(x_i-\bar{x})^2}{n-1}$$

for then we shall have $E(s^2) = \sigma^2$ and therefore s^2 is unbiased. It is possible to show, though we shall not do so, that s^2 is also a consistent estimate of σ^2.

In future we will use s^2 as our estimate of σ^2. The difference between using this and S^2 will be important only if n is small; otherwise the two will be approximately equal. The number $n-1$ which is the denominator of s^2 is often known as the **degrees of freedom** of the sample, an expression which we shall meet frequently in later chapters. Here we are using *one* degree of freedom less than the sample size n, as we have already made *one* estimate, that of $\bar{x}$ for μ.

We therefore use $s = \sqrt{\dfrac{\Sigma(x_i-\bar{x})^2}{n-1}}$ as our estimate of σ. Although it turns out that s itself is not an unbiased estimate it is the unbiasedness of s^2 which is crucial in later work.

With this estimate of σ we are now in a position to know, at least roughly, how accurate our estimate of μ is likely to be.

EXERCISE 11A

1. Boxes of a certain kind of match are sold in packets of ten. The boxes are said to have an average content of 79 matches. 150 packets of ten boxes were examined and the number of boxes in each packet with less than 76 matches counted with the following results:

No. of boxes	0	1	2	3	4	5 or more
No. of packets	70	44	25	8	3	0

Estimate the proportion of boxes of this kind of match which contain less than 76 matches. Also give an estimate of the standard error of your estimate.

2. The number of accidents in a factory for the 30 days of a month were:

No. of accidents	0	1	2	3	4	5	6 or more
No. of days	5	7	8	5	3	2	0

Assuming that the number of accidents has the $\mathscr{P}(\mu)$ distribution, what would you consider to be a sensible estimate for μ? Obtain the value of the estimate in this case and show that in general your estimate is both unbiased and consistent.

3. Tabulate the probability distribution for the sum of the scores of 3 unbiased dice thrown together and verify the results given in §11.3 for the mean and variance of this sum.

4. Show that if x_1 and x_2 are always identical, $\text{Var}(x_1 + x_2) \neq \text{Var}(x_1) + \text{Var}(x_2)$.

5. Write down expressions for $E(x_1 - x_2)$ and $\text{Var}(x_1 - x_2)$ (x_1 and x_2 independent). Confirm your result by writing down the probability function for the difference between the scores on two dice and evaluating the mean and variance of the difference from this function.

6. Three dice are coloured white, red and blue respectively. After casting them a boy scores in the following way. To the white number he adds twice the red number and then subtracts the blue number. Assuming the boy casts the dice a large number of times, calculate the mean and variance of the score.

(J.M.B.)

7. A random variable x takes the value 1 with probability p, and the value 0 with probability $q = 1 - p$. Find the mean and variance of x. Show how you can use these:
   (i) to derive the mean and variance of $B(n, p)$ and
   (ii) to show that the normal approximation to the binomial distribution is a special case of the Central Limit Theorem. (In each case consider n independent variables each with the same distribution as x.)

8. Find the expected mean and variance of the mean of a sample of n values drawn from a table of random numbers. You should carry out the following experiment for various values of n. Draw 100 samples of size n from a random number table.
   (i) Calculate the 100 sample means.
   (ii) Plot a histogram of the distribution of the sample means. Compare the distributions for differing values of n.
   (iii) Estimate the expected mean of the whole population of sample means and compare it with the theoretical value which you calculated earlier.
   (iv) Estimate the variance of the whole population of sample means and again compare it with the theoretical value.
   (If you prefer, you may perform the experiment by tossing a die rather than using a random number table.)

9. A random sample of six values 4·10, 3·96, 4·26, 4·08, 3·74, 3·98 is taken from a population which has mean μ and variance σ^2. Estimate:
   (i) μ, (ii) σ^2, (iii) the standard error of your estimate of μ.

10. The hardness of a plastic material was determined by measuring the indentation produced by a loaded penetrometer. The following measurements were obtained: 4·69, 5·18, 5·36, 4·84, 4·47, 4·86, 4·52, 5·11, 5·01, 4·77. Estimate the size of sample required in order that in future the standard error of the mean will be just less than 0·05. (M.E.I.)

11. Packets of butter have a mean mass of 200 g, and a standard deviation of 1 g. If a sample of 5 packets of this butter is weighed, find the probability that the total mass lies between 995 and 1005 g. Assume that the mass distribution is normal.

12. A sample of 13 observations is taken from a population $\{X\}$ and a sample of 10 observations from an independent population $\{Y\}$. Estimate the variances of the two populations from the following values, and hence calculate the estimated standard errors of $\overline{X}$ and $\overline{Y}$, the means of the sample values of X

and Y respectively. Deduce an estimate for the standard error of $(\bar{X}-2\bar{Y})$.

x 23·3, 43·0, 29·2, 27·5, 41·9, 24·3, 40·7, 34·3, 31·3, 18·2, 30·6, 25·9, 31·7.

y 14·4,   3·3, 11·8, '4·4,   5·2,   7·8,   4·5,   5·9,   8·5, 10·7

(M.E.I.)

13. A population consists of five numbers 1, 3, 4, 7, 10. All possible samples of size two are taken with replacement (i.e. (3, 3) is a possible sample and (3, 4) is different from (4, 3)). Find:
   (i) the expected value of a number chosen at random from the population,
   (ii) the variance of a number chosen at random from the population,
   (iii) the expected value of the sample mean of a sample chosen at random,
   (iv) the variance of the sample mean of a sample chosen at random,
   (v) the expected value of the sample variance S^2 of a sample chosen at random;
   all by direct calculation and verify the results in the chapter.

14. A certain brand of tea contains a picture card in every packet. There are 50 cards in the series and each packet of tea is equally likely to contain any card. If I have already collected 10 different cards, find the probability that I have to buy r more packets of tea before I get the 11th card, and hence find the mean and variance of this number. Generalize the result to find the mean and variance of the number I have to buy for the ith different card ($i = 1, 2, \ldots, 50$). Hence show that the mean and variance of the total number of packets I need to buy to complete the set are:

$$\mu = 50(1+\tfrac{1}{2}+\tfrac{1}{3}+ \ldots \tfrac{1}{50}) \text{ and } \sigma^2 = 50^2(1+\tfrac{1}{2^2}+\tfrac{1}{3^2}+ \ldots \tfrac{1}{50^2})-\mu.$$

15. A variable x has mean μ and variance σ^2. From a set of m independent observations of x, μ is estimated by $\bar{x}$ and σ^2 by s^2. From a similar set of n observations the estimates are $\bar{y}$ and t^2. Show that:

$$\hat{\mu} = \frac{m\bar{x}+n\bar{y}}{(m+n)} \text{ and } \hat{\sigma}^2 = \frac{(m-1)s^2+(n-1)t^2}{m+n-2}$$

are also unbiased estimates.

16. Independent observations x_1, x_2 are taken of the variable x, which has expectation θ. Obtain constants a_1, a_2 so that $X = a_1x_1+a_2x_2$ has expectation θ and variance as small as possible. State the generalization of this estimate to the case of more than two observations. Five observations of x average 3·6 with estimated variance (of the average) 1·5; a further seven observations gave average 6·0 with estimated variance (of the average) 1·0. Estimate the expectation of x and give the estimated variance of your estimate. (Cam.)

17. In an experiment 5 independent observations are made of a quantity x, which has mean value θ and standard deviation σ. At the same time 4 independent observations are made of a quantity y which has mean value 2θ and standard deviation 2σ. If $\bar{x}$ and $\bar{y}$ are the respective sample means, find constants a and b so that $\hat{\theta} = ax+by$ is an unbiased estimate of θ and has variance as small as possible.

18. A bag contains 50 discs of which an unknown number M are coloured red. (It may be assumed that $2 \leqslant M \leqslant 48$.) Two discs are drawn simultaneously and the number r of red ones counted. It is desired to estimate M by cr, where c is a constant chosen so that the estimate is unbiased. Find c and the variance of the estimate.

19. Leaves of a plant are infested by a certain type of insect. The numbers of insects on different leaves are independent and have a Poisson distribution with mean μ. N leaves, with insects on each of them, are collected and it is found that n_r

leaves have r insects on them, $r = 1, 2, \ldots,$ where $N = \sum_{r=1}^{\infty} n_r$.
Since leaves with no insects are not collected, the probability of finding r insects on a collected leaf is no longer the Poisson probability p_r, but becomes modified to kp_r, where k is a constant independent of r. Show that $k = (1 - e^{-\mu})^{-1}$. If $\bar{x}$ is the sample mean number of insects per leaf, derive the expected value of $\bar{x}$. Show that $\bar{x} - (n_1/N)$ is an unbiased estimator of μ.

(J.M.B.)

20. A penny (probability of a head p) is tossed until a head first appears. If this happens on the rth toss, then, since $E(r) = 1/p$, it would seem "sensible" to estimate p by $\hat{p} = 1/r$. Find an expression for the bias of $\hat{p}$, $(E(\hat{p}) - p)$ and show that $\hat{p}$ is not an unbiased estimate. Sketch a graph of the bias.

11.7. A worked example

In section 5 we stated the result that if $x_1, x_2 \ldots x_n$ were a random sample from a normal distribution with mean μ and variance σ^2 then their sample mean $\bar{x}$ was also normally distributed. This is in fact just a special example of a more general result which we will now state.

If $x_1, x_2, \ldots, x_n$ are independent and are all normally distributed and x_i has mean μ_i and variance σ_i^2, then $c_1x_1 + c_2x_2 + \ldots + c_nx_n$, where all the c_i are constants, is also normally distributed. (Notice that if we put each $c_i = 1/n$, each $\mu_i = \mu$ and each $\sigma_i^2 = \sigma^2$, then we have the earlier result.) It is easy to work out the mean and variance of x for we have:

$$E(x) = E(c_1x_1 + c_2x_2 + \ldots + c_nx_n)$$
$$= E(c_1x_1) + E(c_2x_2) + \ldots + E(c_nx_n)$$
$$= c_1E(x_1) + c_2E(x_2) + \ldots + c_nE(x_n)$$
$$= c_1\mu_1 + c_2\mu_2 + \ldots + c_n\mu_n.$$
$$\mathrm{Var}(x) = \mathrm{Var}(c_1x_1 + c_2x_2 + \ldots + c_nx_n)$$
$$= \mathrm{Var}(c_1x_1) + \mathrm{Var}(c_2x_2) + \ldots + \mathrm{Var}(c_nx_n)$$
$$= c_1^2\mathrm{Var}(x_1) + c_2^2\mathrm{Var}(x_2) + \ldots + c_n^2\mathrm{Var}(x_n)$$
$$= c_1^2\sigma_1^2 + c_2^2\sigma_2^2 + \ldots + c_n^2\sigma_n^2.$$

We can best show the use of this result with an example.

George is travelling from Canterbury to Birmingham by car. The journey consists of four stages: (1) Canterbury to the outskirts of London, (2) crossing London to reach the M1, (3) along the M1 to the outskirts of Birmingham, (4) reaching his destination in Birmingham. He has done the journey many times and has found that the times for the various sections of the journey are approximately normally distributed with the means and standard deviations given in the following table:

stage	mean (minutes)	standard deviation (minutes)
1	70	10
2	60	10
3	65	5
4	30	5

Assuming that the time for each section of the journey is unaffected by all the other times, what is the probability that George will complete the journey in less than four hours, and what is the probability that George spends longer crossing London than on the M1?

If we let x_1 be the time for the first section, x_2 for the second section, and so on, then the total journey time is given by $x = x_1 + x_2 + x_3 + x_4$. By the theorem above, x will be normally distributed with

$$E(x) = E(x_1) + E(x_2) + E(x_3) + E(x_4) = 70 + 60 + 65 + 30 = 225$$

and $\text{Var}(x) = \text{Var}(x_1) + \text{Var}(x_2) + \text{Var}(x_3) + \text{Var}(x_4)$

$$= 10^2 + 10^2 + 5^2 + 5^2 = 250.$$

P(George's journey time is less than 4 hours) $= P(x \leqslant 240)$

$$= \Phi\left(\frac{240 - 225}{\sqrt{250}}\right) = \Phi\left(\frac{3\sqrt{10}}{10}\right) = \Phi(0 \cdot 95) \simeq 0 \cdot 83.$$

Hence the required probability is $0 \cdot 83$.

For the second part we need to know the probability that x_2 is greater than x_3.

$$P(x_2 > x_3) = P(x_2 - x_3 > 0).$$

If we let $y = x_2 - x_3$ then y will also be normally distributed.

$$E(y) = E(x_2 - x_3) = E(x_2) - E(x_3) = 60 - 65 = -5,$$
$$\text{Var}(y) = \text{Var}(x_2) + \text{Var}(-x_3) = \text{Var}(x_2) + (-1)^2 \text{Var}(x_3)$$
$$= 10^2 + 5^2 = 125.$$

$$P(y > 0) = 1 - P(y \leqslant 0) = 1 - \Phi\left(\frac{0 - (-5)}{5\sqrt{5}}\right)$$

$$= 1 - \Phi(0 \cdot 45) \simeq 1 - 0 \cdot 67 = 0 \cdot 33.$$

Hence the probability that he spends longer crossing London than on the M1 is $0 \cdot 33$.

EXERCISE 11B

1. A man travels from his London office to his home by a tube journey from station A to station B. His walking times to A and from B add up to 5 minutes with negligible variation, the variable factors in the journey being as follows, measured in minutes:

	Mean time	Standard deviation
(i) waiting for train	8	2·6
(ii) train journey	47	1·8

Assuming that these two factors are independent and normally distributed, find the mean and standard deviation of his whole journey. Estimate the probability of the *whole* journey taking:

(a) less than 52 minutes,

(b) more than 65 minutes,

(c) between 57 and 62 minutes.   (M.E.I.)

2. The distribution of breaking loads for strands of rope is approximately normal, with mean 20 units and standard deviation 2 units. A rope is assumed to be made up of 64 independent strands, and to have a breaking load which is the sum of the breaking loads of all the strands in it. Find the probability that such a rope will support a load of 1300 units. The manufacturers wish to quote a breaking load for such ropes that will be satisfied for 99% of the ropes. Determine what breaking load should be quoted. (J.M.B.)

3. The total weight of food delivered by a filling machine varies normally with mean 720 g and standard deviation 6 g. The cans that are being filled are of weight normally distributed with mean 238·5 g and standard deviation 4·5 g. Show that the standard deviation of the weight of a filled can will be 7·5 g, and estimate the percentage of filled cans that will weigh less than the advertised weight of 960 g. What should be the mean weight delivered by the filling machine so that only one per cent of the filled cans weigh less than the advertised weight? (M.E.I.)

4. The heights of boys of a certain age group are normally distributed with mean 150 cm and standard deviation 5 cm; the heights of girls of the same age are normal with mean 147 cm and standard deviation 4·5 cm. Find the probability of differences in height greater than 5 cm between:

(a) two boys of this age,
(b) two girls,
(c) one boy and one girl. (Cam.)

5. The diameter of a mass-produced rod is a normal variable with mean 0·85 cm and standard deviation 0·03 cm. The internal diameter of the socket which holds the rod is a normal variable with mean and standard deviation respectively 0·94 cm and 0·04 cm. Rods and sockets are paired randomly during assembly. What proportion of assemblies:

(i) are rejected because the socket is too small for the rod,
(ii) are rejected because the rod is too loose in the socket, which is the case if the diameter of the socket exceeds that of the rod by more than 0·17 cm?

If the diameter of the socket exceeds that of the rod by more than 0·12 cm, the assembly requires a spring washer. What proportion of accepted assemblies require a spring washer? (Cam.)

6. Three observations $\{x_i\}$ are taken from a normal distribution with mean 1 and standard deviation 3, and twenty-five observations $\{y_i\}$ are taken from a normal distribution with mean 4 and standard deviation 4. Find

$$P\{\bar{x} \leqslant 1·25\bar{y} - 2\} \text{ and } P\{\bar{x} \geqslant 2 - 1·25\bar{y}\}.$$

Show by means of a diagram how these probabilities are related to

$$P\{|\bar{x}| \leqslant 1·25\bar{y} - 2\}.$$

(M.E.I.)

Conclusions from Sample Data I

12.1. Hypothesis testing; the binomial distribution with small samples

We must now consider how the theoretical distributions we have been looking at in Chapters 7–10 help us in real-life situations. No actual set of data is likely to correspond exactly to what is predicted by a theoretical model. However, a model is very important where we wish to draw conclusions about a whole population. One example would be trying to decide which party will win an election, and by what margin, when the opinion poll necessarily only collects the views of a sample.

Another, which we will consider in detail, might face a committee that makes grants towards medical research. A doctor requests financial backing for the study of a drug. This drug is alleged to give a significant improvement in the recovery rate of patients suffering from a fatal disease. The doctor bases his claim on the fact that, of the 15 patients treated with the drug, 8 recovered, whereas the chances of survival had previously been found from long experience to be about $\frac{1}{3}$. Is there enough evidence of the effectiveness of the drug to warrant large-scale expenditure on an intensified research programme?

Obviously it is extremely difficult to make an important decision on the basis of such a small sample; but sometimes, and especially often in medical work, it may be both risky and difficult to obtain a larger one. So, accepting the size of the sample as 15, we are forced to decide how many would need to recover before we accept it as a significant improvement on the old situation, in which the expected number of recoveries was $\frac{1}{3}$ of 15, which is 5.

If 15, 14, or even 13 survived, we would have little hesitation in acclaiming the drug. If no more than 6 recovered, we would probably decide that there was insufficient evidence to show that the new drug was superior to the old one. Somewhere between 6 and 13 then we must draw a line, so that anything to the left of it is labelled "not significantly different from expected", and anything to the right of it we would agree to call "significant". We could of course place the line by some sort of hunch. But how could we be sure that our decision would then be the

164

same as that of the other committee members or that, if another case came up with a sample of 25 patients, we could draw a similar line which would be fair to both cases?

We clearly need some objective criterion of significance. If we look closer at this situation, we see that the 15 patients can be seen as 15 trials, each of which succeeds or fails, and the outcome of the 15 can reasonably be assumed to be independent. If we assume that each trial has the same probability p of success, we are in a binomial situation, and we might use $B(15, p)$ as a model of it, for some value of p.

The problem we are really faced with is deciding between two rival hypotheses which give differing values for the parameter p, namely $H_0 : p = \frac{1}{3}$ and $H_1 : p > \frac{1}{3}$.

The first hypothesis H_0 is called the *null hypothesis* because it assumes the drug has no effect and therefore the survival rate is the same as before. It is the cautious conservative hypothesis that alleges that any apparent improvement in the survival rate is simply the result of a lucky sample. The doctor wishes only to show that his drug causes an improvement, and hence his hypothesis H_1, known as the *alternative hypothesis*, is that p has increased. It could be argued that H_0 should really be $p \leqslant \frac{1}{3}$, since it is possible that the new drug is actually worse than the old. However in taking $H_0 : p = \frac{1}{3}$ we are setting up the strongest test for the new drug. In general we try to ensure that the null hypothesis predicts a definite value for the parameter in question, whereas the alternative suggests a range of values.

In deciding whether the evidence favours H_0 or H_1, we are playing the part of jurors, who, even though they may suspect that the defendant is guilty, are bound to return a verdict of "not guilty", or more properly "guilt not proven", unless they are satisfied "beyond all reasonable doubt" that the evidence contradicts this. We must also take the cautious view that the survival rate has not changed, unless the data show this to be improbable. As p is a population variable, in this case the probability that anyone with the disease will survive, we can never prove, without access to the whole population, that it takes, or does not take a certain value. For instance, even if all 15 patients survived, this does not definitely contradict H_0; for with $p = \frac{1}{3}$, there is still a probability of $(\frac{1}{3})^{15}$ of this happening. However this probability is so small that we are likely to discount its occurrence, and to reject H_0, while acknowledging that we could be wrong if the very unlikely were to occur.

Coming back to the case of our doctor and his drug, we have still to decide exactly where to draw the line between accepting and rejecting the null hypothesis. Assuming the null hypothesis to be true, the number

x of patients in the sample who survive will have the $B(15, \frac{1}{3})$ distribution, and we can tabulate the probability function as below correct to 3 decimal places.

x	0	1	2	3	4	5	6	7	8	9
$P(x)$	0·002	0·017	0·059	0·130	0·195	0·214	0·178	0·115	0·057	0·022

x	10	11	12	13	14	15
$P(x)$	0·007	0·002	0·000	0·000	0·000	0·000

As we suspected, if $p = \frac{1}{3}$, any of the values 11–15 are very unlikely, so we might agree to draw the line between 10 and 11 and reject the null hypothesis whenever x was greater than or equal to 11.

Plan 1. Reject H_0 if $x \geqslant 11$.

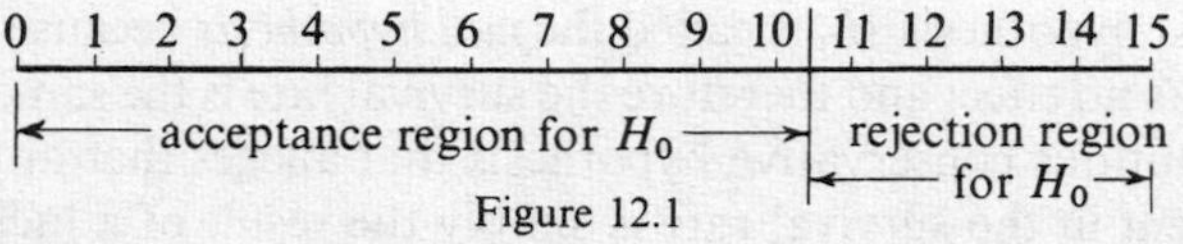

Figure 12.1

For $p = \frac{1}{3}$ we have $P(x \geqslant 11) = P(11) + P(12) + P(13) + P(14) + P(15)$

$\simeq 0.002$. (Rounding errors in these results are negligible.)

Hence the probability that x falls in the rejection region if H_0 is true is roughly 0·002. We would therefore only wrongly reject H_0 with probability 0·002 or 0·2%. This probability is called the **significance level**.

Although Plan 1 has the advantage of making the probability of wrongly rejecting H_0 very small, it is rather unfair to the doctor, as it accepts the *status quo* unless $x \geqslant 11$. In other words at least 11 out of the 15 must survive before we will consider that the drug might be having some effect. It looks as if we would be likely to accept the null hypothesis when in fact it was false. Thus a numerically small significance level is like a jury who only reject the "innocence" hypothesis if the evidence is almost incontrovertible. They may not convict many innocent men, but on the other hand they will let many guilty ones go free.

Suppose we swing the other way and agree to accept that the drug was effective provided more than 6 survived.

Plan 2. Reject H_0 if $x \geqslant 7$.

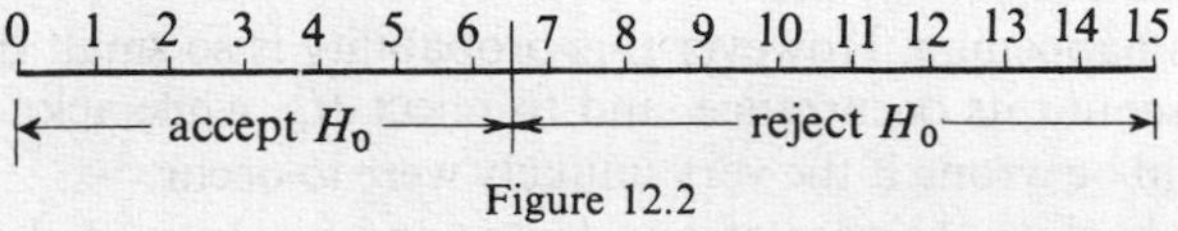

Figure 12.2

But for $p = \frac{1}{3}$, $P(x \geqslant 7) = P(7) + P(8) + P(9) + P(10) + P(11) + P(12) + P(13) + P(14) + P(15) \simeq 0.203$.

So with probability about 0·2, in other words in roughly one sample out of every 5, 7 or more in a sample of 15 would recover even though the general recovery rate was still $\frac{1}{3}$. This means that with Plan 2, we are so keen to accept the value of the drug that we run a 20% chance of failing to reject a worthless drug presented to us. Hence taking a numerically high significance level of 20%, we are being too ready to reject the null hypothesis, which is rather like a jury which is so ready to convict a prisoner that in its efforts not to let any guilty men go free it wrongly convicts 20% of the innocent.

Just as the jury must find an acceptable compromise between being too lenient and being too severe, we must choose a significance level which is reasonably fair to both sides.

In practice we usually choose 5% as a suitable significance level, which means that the probability of wrongly rejecting H_0 is 0·05, and hence there is a 1 in 20 chance that we shall give the doctor the benefit of the doubt when he does not deserve it. Sometimes, if we are especially anxious to retain the null hypothesis, we use a significance level of 1% or 0·1% instead, or conversely if we are anxious to favour the alternative hypothesis we might choose a level of 10%. In order to be quite fair, we should always determine what level of significance we are going to use *before* we know the actual experimental results, so that they do not affect our choice.

So we are led to:

Plan 3. Reject H_0 if $x \geqslant k$, where k is to be chosen so that $P(x \geqslant k) = 0·05$. (Of course, since we are dealing with a discrete variable, we are most unlikely to be able to choose a k such that $P(x \geqslant k) = 0·05$ but choosing k to be the smallest integer so that $P(x \geqslant k) \leqslant 0·05$ will give us essentially the same significance level.)

We know $P(x \geqslant 11) = 0·002$, $P(x \geqslant 10) = 0·009$, $P(x \geqslant 9) = 0·031$, $P(x \geqslant 8) = 0·088$. So we see that if we choose $k = 9$ we have the required conditions since

$$P(x \geqslant 9) \leqslant 0·05 \text{ and } P(x \geqslant 8) > 0·05.$$

Hence we agree to reject H_0 if $x \geqslant 9$.

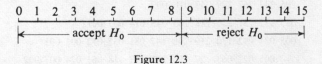

Figure 12.3

As we were given that $x = 8$, which does not lie in the rejection region for H_0, we must regretfully say to the doctor that the beneficial effect of the drug is "not proven" and that the evidence is not sufficient to shake our opinion that the recovery rate is $\frac{1}{3}$.

12.2. Confidence intervals: the binomial distribution for a small sample

In the last example we wanted to check whether a particular value of the population proportion, namely $p = \frac{1}{3}$, was compatible with the results we obtained in a sample. Sometimes, however, we wish to use our sample data to solve the more general problem of finding the range in which the population value is most likely to lie.

A quality control inspector at a small jam factory selects 15 jars of jam each day and examines them to make sure that the proportions of various ingredients are as stated on the jar. If, on one particular day, he finds that 2 are definitely substandard, what would he estimate as the maximum proportion of that day's production which is also substandard?

As in the medical example, it would be extremely naive to say that as $\frac{2}{15}$ of his sample was substandard, we would predict that $\frac{2}{15}$ of the day's production was as well. Although this is certainly the best single or "point" estimate of the proportion, it tells us nothing about how high the real value might be.

We are again in a binomial situation, each jar selected being a trial, provided we agree to assume independence of the jars. We shall therefore take the model for the number of "failed" jars to be $B(15, p)$ where p is the proportion of substandard jars in the population. What we need to find is the highest value of p compatible with our having found 2 substandard jars out of the 15 in the sample.

To do this, we consider the probable outcomes of the sampling test for various values of p. The five rows of the following table give the probabilities of obtaining each different number of substandard jars out of the 15 in the sample for the five values of $p: p = 0.1, 0.2, 0.3, 0.4, 0.5$. (Entries are given to 3 decimal places and all missing values are equal to 0.000 to 3 decimal places.)

No. of defective jars in sample of 15

	0	1	2	3	4	5	6	7	8	9	10	11	12	13	14	15
$p = 0.1$	·206	·343	·267	·129	·043	·010	·002									
$p = 0.2$	·035	·132	·231	·250	·188	·103	·043	·014	·003	·001						
$p = 0.3$	·005	·031	·092	·170	·219	·206	·147	·081	·035	·012	·003	·001				
$p = 0.4$		·005	·022	·063	·127	·186	·207	·177	·118	·061	·024	·007	·002			
$p = 0.5$			·003	·014	·042	·092	·153	·196	·196	·153	·092	·042	·014	·003		

Let us take for example the case $p = 0.2$. Corresponding to this value we frame the null hypothesis $H_0 : p = 0.2$, and the alternative hypothesis $H_1 : p < 0.2$. We can draw up the acceptance region for the sample result for a 5% level of significance using the second row of the table. As in the medical example, we arrange the acceptance region so that the probability of x lying outside it is as large as possible, while still being less than 0·05. The only difference is that this time H_1 gives $p < 0·2$, so that we choose the rejection region to include only values at the *lower* end of the scale. This arrangement enables us to see for which values of p our sample value of 2 out of 15 lies in the rejection region.

$$P(x \leqslant 0) = 0·035 \leqslant 0·05, \quad P(x \leqslant 1) = 0·035 + 0·132 = 0·167 > 0·05$$

Thus any sample value greater than 0 leads to the acceptance of the hypothesis $p = 0·2$. In particular the sample value we obtained of 2 out of 15 is compatible with $p = 0·2$.

$H_0 : p = 0·2$
$H_1 : p < 0·2$
$x = 2$, so accept H_0.

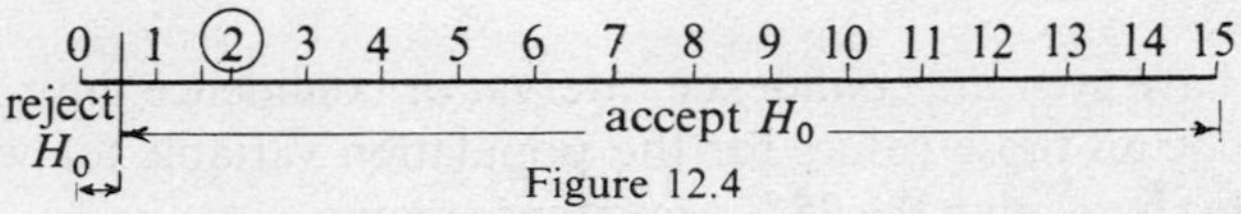

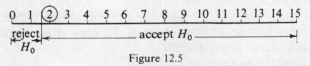

Figure 12.4

As we are looking for the largest possible value of the population proportion p which allows our sample value of 2, we can now proceed to test $p = 0·3$ and $p = 0·4$ in a similar way.

$H_0 : p = 0·3$
$H_1 : p < 0·3$
$x = 2$, so accept H_0.

Figure 12.5

$H_0 : p = 0·4$
$H_1 : p < 0·4$
$x = 2$, so reject H_0.

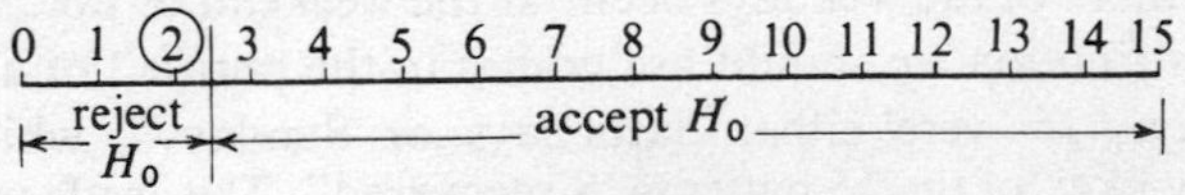

Figure 12.6

We must therefore reject the hypothesis $p = 0·4$ on our sample result of 2, while accepting the possibility that p might equal 0·3. There is clearly no need to test higher values of p, for if $p = 0·4$ is unacceptable, so will be $p = 0·5$ and so on.

Hence the largest value of the population proportion of substandard jam-jars that is compatible with our sample figure, using a 5% level of significance, is somewhere between 0·3 and 0·4. A more accurate estimate could be obtained by trying out other values of p in this range. It is therefore not very comforting for the quality inspector to find that although only $\frac{2}{15} \simeq 0.133$ of his sample of jars was defective, this may indicate that the population proportion is as high as 0·3 or more, though it is not likely to reach 0·4.

In this example we have not simply been checking one value for the population proportion. Instead we tested a number of values to establish the range of possible values, in this case with the approximate result $0 < p < 0.4$. This range is known as 95% **confidence interval** for p, and the procedure is known as **interval estimation.** More strictly, as we are not concerned with giving a lower limit to the interval in this particular example, 0·4 is the *upper 95% confidence limit* for the population proportion.

As we have seen the confidence interval or confidence limit is drawn up to include all those values for the population variable for which the sample value lies within the 95% acceptance region.

12.3. Hypothesis testing: the binomial distribution with a large sample

"I haven't had a round of golf for ages," said George. "Why does it always rain at the weekend?"

"It doesn't," I replied, "you just don't notice it on the other days." George came back jubilant a few days later.

"I was right," he said, "I rang the Meteorological Office and they said that, of a sample of 40 wet days, 15 were either Saturdays or Sundays; and that is nearly half!"

"That doesn't prove anything," I said.
Who was right?

Again we have to make a decision between the two rival hypotheses, that either more of the wet days occur at the weekend or not. The hint as to what distribution we should use comes in the phrase "of a sample of 40 wet days, 15 were either Saturdays or Sundays", which sounds remarkably like "of the 15 patients, 8 recovered". This leads us to expect that a binomial model might fit. The number of trials must be the total number of wet days, 40, with "success" if one of them happens to be a Saturday or a Sunday. There are obvious complications about how the sample is chosen. Strictly it should be a random sample of all wet days—not, for example, the last 40 wet days. We will ignore these complications, while realizing that any results we derive may be subject to error for this reason. The number of weekend days in the sample will then have the distribution $B(40, p)$ where p is the probability that a wet day is a Saturday or Sunday.

George's hypothesis is that more wet days occur at the weekend than we would expect. Will this do for the null hypothesis? Clearly not, because it does not give us a definite value for p, just that p is greater than some expected value. Also, the null hypothesis is normally the cautious one, that everything occurs as expected, whereas George appears to be postulating some supernatural intervention which ensures more rain at weekends.

But what is the expected value of p? Remembering that p is the probability that a wet day is a weekend day and two days in every seven are weekend days, we would expect that $p = \frac{2}{7}$. We therefore have the null hypothesis $H_0 : p = \frac{2}{7}$ while George's claim becomes the alternative hypothesis $H_1 : p > \frac{2}{7}$. (It is probably worth while at this point checking that the mean number of weekend days out of 40 predicted by H_0 would be $\frac{2}{7} \times 40 \simeq 11$, for if the actual number had been less than 11 George would clearly have had no case at all.)

We shall again choose a 5% significance level, and we could go through the process of working out $P(40)$, $P(39)$, ... and so on until we arrived at a 5% total to give us the rejection region. However we can save a lot of trouble by remembering that we discovered in Chapter 10 that if the size of a binomial sample was large enough, then the normal curve with the same mean and variance gives approximately the same shape of frequency curve; therefore we can approximate to $B(n, p)$ by $N(np, npq)$. In this case, if H_0 is true $p = \frac{2}{7}$ and the distribution will be approximately $N(40 \times \frac{2}{7}, 40 \times \frac{2}{7} \times \frac{5}{7})$ or $N(\frac{80}{7}, \frac{400}{49})$.

This means that the boundary of the 5% region will be much easier to find.

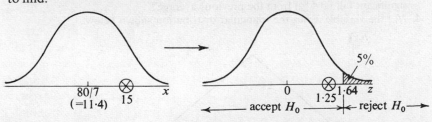

Figure 12.7   Figure 12.8

All we need to do is to look up in the normal distribution tables (or table 2) for the z-value to the left of which 95% of the area lies. (Compare this with the snake example in Chapter 10.)

We find in this case that $z = 1.64$, so we are ready to go ahead.
$H_0 : p = \frac{2}{7}, H_1 : p > \frac{2}{7}$.
Reject H_0 if $z > 1.64$.
But $z = (x - \mu)/\sigma = (x - np)/\sqrt{(npq)} = (15 - \frac{80}{7})/\frac{20}{7} = \frac{25}{20} = 1.25 < 1.64$
Hence we would accept H_0 at the 5% significance level and the result is not significant. (Strictly, as we saw in Chapter 10, since we are approximating to a discrete variable by a continuous one, we should substitute 14·5 rather than just 15 in the expression for z. This however would not alter our conclusion as it would give $z = 1.1 < 1.64$.) In fact such corrections always reduce the significance of results.

We must therefore conclude that the evidence that 15 out of 40 wet days were either Saturdays or Sundays is not enough to suggest that these days are wetter than any others, and George's hypothesis is therefore "not proven".

EXERCISE 12A

1. Five people taken at random are each asked to taste a sample of Brand A and a sample of Brand B margarine. Four of them prefer Brand A. Do you feel that this provides conclusive evidence that people "prefer Brand A to Brand B"? Give your reason.   (S.M.P.)

2. At a seed-testing station it is found that a proportion 0·4 of a certain type of seed is fertile. By accident the remaining stock of this seed (whose total amount is very large) is completely mixed with an equal quantity of a second type of seed which is believed to be completely infertile. If this latter assumption is true, what is the probability that a seed taken at random from the mixture will germinate? Each of seven pots is planted with two seeds taken at random from the mixture. Six pots eventually produce one or more plants each. Is this result consistent, at the 5% level of significance, with the infertility postulated for the second type of seed?   (Cam.)

3. A television shop sells on average 6 colour televisions a week. In the week following a price rise, the number drops to 2. Would we be justified in saying that the price rise had caused a significant drop in sales? What is the greatest number of sets which could be sold in that week which would represent a significant fall (at 5%) from the previous average?

4. H_0: the variable X has the triangular distribution shown below:

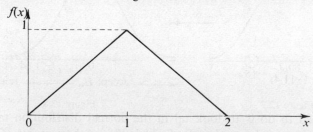

H_1: the variable X has a rectangular distribution on the interval $(-1, 1)$.
H_1^*: the variable X has a triangular distribution similar to that in H_0 but

whose mean is different from 1. Construct significance tests to decide between (a) H_0 and H_1; (b) H_0 and H_1^* on the evidence of a single observation x of the variable X such that the probability of accepting H_1 (or H_1^*) when H_0 is true is 0·05. Calculate also the probability of accepting H_0 when H_1 is true and say whether your test for discriminating between H_0 and H_1 is a good one. How would you improve it?

5. What do you think is the probability that any person chosen at random is left-handed? Formulate a hypothesis giving a value of p correct to 1 decimal place, and use the table on page 168, to decide on your acceptance and rejection regions for a sample of 15 people. Test your hypothesis by choosing 15 people at random and seeing whether the number who are left-handed supports it. You can of course use other characteristics or subjects, e.g. the proportion of cars of a certain make, the proportion of people wearing glasses, a pink shirt, etc., provided the sample size is the same.

6. The proportion of defective articles produced in a certain manufacturing process has been found from long experience to be 0·1. In the first batch of 50 articles produced by a new process 3 were defective. Using the normal approximation to the binomial distribution, calculate the probability of so small a number of defective articles if the proportion of defective articles is unchanged, and explain how to use this probability in a significance test of the null hypothesis that the new process is no better than the old. (Cam.)

7. Over a number of years an average of 40 out of every 100 patients who underwent a difficult operation survived. Last year new medical techniques were introduced and 73 out of 150 patients survived the operation. Explain whether or not the maintaining of these techniques is statistically justified.

(S.M.P.)

8. The proportion of blue-eyed persons in a certain large population is 0·2. A group of ten persons is selected at random. Calculate the probability that the number of blue-eyed persons in the sample is:
   (i) exactly 3,
   (ii) at least 3.
A second sample of ten persons is examined; calculate the probability that the total number of blue-eyed persons in the two samples combined is exactly 2. A random sample of 200 persons from a second large population is examined for eye colour. In this sample 50 persons are found to have blue eyes. Is this result consistent with the hypothesis that the proportion of blue-eyed persons in this second population is 0·2? What would have been your conclusion if there had been 200 blue-eyed persons in a random sample of 800 persons from this second population? (M.E.I.)

9. The sex ratio at birth in domestic cattle is approximately 105 males to 100 females. In a sample of wild cattle 58 bull calves were obtained in 100 births. Is this sample sufficient to show at the 5% level of significance that the sex ratio in wild cattle is different from that in domestic cattle? In a further sample of 125 single births, 76 bull calves were obtained. What conclusion may be drawn from the two samples together? (Cam.)

10. It is claimed that 90% of women use Froth soap powder. In a random survey of 400 homes, 60 women said that they did not use Froth. Investigate whether the claim is justified. (J.M.B.)

11. In an investigation of preferences between pre-packed and fresh-cut bacon, 69 housewives preferred pre-packed and 52 preferred fresh-cut. Does this result provide evidence (at the 5% level of significance) of a difference in acceptability? Find those values of the percentage of housewives preferring pre-packed which could just be regarded as reasonable, that is which just do not differ from the observed proportion at the 5% level of significance.

(Cam.)

12. In an opinion poll carried out before a local election, 502 people out of a random sample of 920, declare that they will vote for a particular one of the two candidates contesting the election. Discuss whether there is significant evidence of a distinct preference in the electorate for one or other of the two candidates. Calculate a symmetric two-sided approximate 99% confidence interval for the proportion of the electorate who would on the basis of this sample be expected to vote for the candidate preferred in the opinion poll.

(J.M.B.)

13. A product is made up in batches of 5 for distribution, and in 90 such batches there is a total of 135 faulty items. Obtain a confidence interval for p, the probability that a single item will be faulty. Deduce a confidence interval for the average number of faulty items in a batch. (You may use the normal approximation.) (J.M.B.)

14. A gambler obtains a coin which is alleged to be biased so that the probability p that it falls heads is 0.3. He wishes to test whether the coin is biased in this way or whether it is unbiased, and decides to do so by tossing the coin 100 times with the intention of accepting it as biased if he gets at most 37 heads. Calculate the Type I error, i.e. the probability that he rejects the coin when, in fact, $p = 0.3$, and also the Type II error, i.e. the probability that he accepts the coin when, in fact, $p = 0.5$. Explain briefly how he could devise a test so that the Type I error does not exceed 1% and the Type II error does not exceed 2%.

(Cam.)

15. In a community with a large number of drivers an average of 100 are involved in accidents each year. In a certain year 120 have accidents: carry out a significance test to discover whether this represents a real increase. (J.M.B.)

12.4. Hypothesis testing: mean from large sample

The local supermarket manager decides to install equipment to produce background music in his supermarket and wants to find out whether this has any significant effect on the amount people spend there. On the first Saturday after its installation he chooses 200 customers at random and finds that they spend an average of £2·78 each with a standard deviation of £1·20, compared with the well-established normal Saturday figure of £2·60. Is he right in assuming that the music has had no real effect and that the difference in mean amount spent is simply due to sampling errors?

This time we are interested in measuring what is, for all practical purposes, a continuous variable; namely, the amount people spend on Saturdays. Furthermore we are concerned with a mean of a sample of size 200; let us call it $\bar{x}$. Since the sample size is large, we are justified in assuming that, whatever the distribution of the amount spent, the mean amount $\bar{x}$ will be approximately normally distributed with mean μ and variance σ^2/n, where μ is the mean amount spent in the whole population, σ^2 is the population variance and n is the sample size. The parameter about which we wish to make our hypotheses is μ. We are not told the value of σ^2, but we are told that the standard deviation of the sample, s, is £1·20. As 200 is a relatively large sample size, we are justified in using the sample standard deviation s as a reasonable approximation for σ and behaving just as if we knew that this was the exact value for σ. (We shall see later that this is not the case when the sample size is small. Then we shall only be justified in using the test described in this section in the rare event of actually knowing the population standard deviation. In other cases a different procedure has to be followed—we shall meet this in Chapter 13.)

So we have that the mean Saturday spending £$\bar{x}$ is distributed approximately as $N(\mu, 1·2^2/200)$.

What about the hypotheses? The null hypothesis which gives a definite value for μ must be that it has remained unchanged at £2·60. Before the experiment took place, it is conceivable that the music could have either increased or decreased the value of μ, so the alternative hypothesis must cater for both possibilities.

So we have $H_0 : \mu = 2·60$
$\qquad\qquad H_1 : \mu > 2·60$ or $\mu < 2·60$, i.e. $\mu \neq 2·60$

It might seem that the alternative $\mu < 2·60$ is unnecessary, as after the experiment we know that the sample mean is £2·78 and therefore it would seem inconceivable that the mean amount had decreased. However this is a judgment we are making in the light of the data, and the hypotheses should always be formulated before the actual results are obtained, or, if this is not possible, should be chosen without being biased by the data. Only if we were definitely convinced before the experiment that the mean amount would not be decreased, would we be justified in omitting $\mu < 2·60$; we prefer to take an open view.

When we are considering alternative hypotheses which are supported by a value of the parameter either greater or smaller than that predicted by H_0, we call the hypothesis test a *two-tailed test*. We must divide up our rejection region so that we reject H_0 for values of μ substantially greater or substantially smaller than the predicted one, and the region therefore covers both "tails" of the distribution. This is a different case from those already discussed, which have in all cases been *one-tailed tests*, with the rejection region all at one end of the distribution.

If we took the same value of z that we used in determining the 5% rejection region in the normal approximation to the binomial distribution, we should have $z = 1.64$. However we should have to add a similar region at the other "tail" of the curve, agreeing to reject H_0 if $z < -1.64$ also.

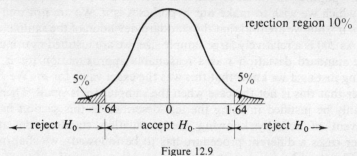

Figure 12.9

We can see that 5% at each end would give us a total rejection region of 10%, and we would be working at 10% and not 5% significance. We must therefore take only $2\frac{1}{2}$% of the area at each end, giving a total of 5% and a probability of 0.05 that z will fall in the rejection region.

If we look in the normal tables for the value of z for which 0.975 of the area is to the left of it, we find $z = 1.96$ (table 2 gives this value directly).

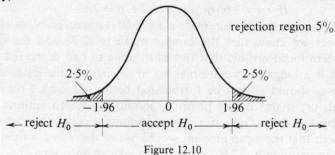

Figure 12.10

Hence for our supermarket example we would reject H_0, that the mean amount spent was £2·60, in favour of H_1, that the mean amount spent was different from £2·60, if either $z > 1.96$ or $z < -1.96$, where

$$z = \frac{\bar{x}-\mu}{s\sqrt{n}}$$

We have $z = \dfrac{2.78-2.60}{1.2/\sqrt{200}} = 2.12 > 1.96.$

So we would reject H_0 and accept H_1. We therefore conclude that the evidence is sufficient (at the 5% level) to prove that the music alters Saturday spending.

12.5. Confidence intervals: mean from large sample

The result of the last section shows that the average amount spent by the sample on Saturdays is compatible with the hypothesis that the mean spending has changed, but we may now want to know what values the mean spending is likely to take. Let us set up a confidence interval for the average amount spent in the light of our sample.

In the binomial case the only way to do this was to try various values for the unknown parameter p and see whether we would reject the null hypothesis for these values of p. For the normal distribution we can carry out the same procedure, but the arithmetic is much easier.

Let us suppose that we wish to test the null hypothesis $\mu = \mu_0$ against the alternative $\mu \neq \mu_0$. We know we would reject H_0 if $\mu_0 = 2 \cdot 60$; now we wish to know for which new values of μ_0 we would accept it. Working as before, we will accept H_0 provided $z < 1 \cdot 96$ and $z > -1 \cdot 96$

where $\quad z = \dfrac{\bar{x} - \mu_0}{s/\sqrt{n}} = \dfrac{2 \cdot 78 - \mu_0}{1 \cdot 2/\sqrt{200}}.$

Hence z lies in the acceptance region provided

$$-1 \cdot 96 < \frac{2 \cdot 78 - \mu_0}{1 \cdot 2} \sqrt{200} < 1 \cdot 96$$

i.e. $\quad \dfrac{-1 \cdot 96 \times 1 \cdot 2}{\sqrt{200}} < 2 \cdot 78 - \mu_0 < \dfrac{1 \cdot 96 \times 1 \cdot 2}{\sqrt{200}}$ or $-0 \cdot 17 < 2 \cdot 78 - \mu_0 < 0 \cdot 17$

$$\text{or } 2 \cdot 61 < \mu_0 < 2 \cdot 95.$$

The range of values of μ_0 for which we would accept H_0, against the two-sided alternative is therefore £2·61–£2·95 (or £2·78±0·17). This is called the 95% confidence interval for the new mean amount spent. Notice that the fact that the 95% confidence interval is £2·61–£2·95 does *not* mean that there is a probability of 0·95 that μ lies within this interval, for μ is a fixed quantity and does not vary. It does mean however that our obtained value of £2·78 lies within the 95% acceptance region for any value of μ within the range £2·61–£2·95.

12.6. Two kinds of error

During the course of this chapter we have been drawing together our calculations of the statistics of a sample and our knowledge of possible probability models of the random variables concerned in order to draw "reasonable" conclusions about the population as a whole. Because statistical decisions are always based on the incomplete evidence provided by a sample, we are never in a position to say that any particular hypothesis is definitely true or false. We can only conclude that the evidence tends to support one hypothesis rather than the other. The great advantage of a statistical procedure over a common-sense one is that the former allows us to say with what probability we are wrong.

For instance, if we choose the rejection region so that if the null hypothesis is true, the likelihood of obtaining a result within the rejection region is less than or equal to 5%, then we are working at the 5% significance level. This means that when the null hypothesis is true, we will mistakenly choose to reject it in 5% of cases on average. The error of wrongly rejecting the null hypothesis is called the *Type I error* and the probability of its occurring must always be equal (or approximately equal in discrete cases) to the significance level at which we are working.

The other mistake we are likely to make is to accept the null hypothesis when the alternative is true. This is known as the *Type II error* and its probability will depend on the alternative hypothesis and is more difficult to work out. It is the probability that the result lies in the acceptance region for the null hypothesis if the alternative is true.

EXERCISE 12B

1. A machine produces ball-bearings whose diameters are normally distributed with mean 2·00 mm. The machine is modified, and a sample of 100 ball-bearings subsequently produced is found to have mean 2·01 mm and standard deviation 0·05 mm. Determine whether there is sufficient evidence that the modification has affected the mean diameter of the ball-bearings.   (J.M.B.)

2. The average escalation of development costs on aircraft projects over the last 10 years has been 120% with a standard deviation of 50 percentage points. Under a new government the first 16 projects show a mean escalation of 150%. If the standard deviation can be assumed to have remained constant, at what levels is the increase a significant one?

3. The breaking strain of a certain type of chain is expected to be 1300 kg and the standard deviation is known to be 40 kg. The mean breaking strain of 9 test lengths was 1323 kg. Was there significant evidence of a change in breaking, and if so at what level of significance? The manufacturing process was slightly modified, the standard deviation remaining unchanged: the first 9 test lengths gave breaking strains (kg) 1254, 1301, 1344, 1318, 1394, 1335, 1262, 1337, 1362. Is there significant evidence that the breaking strain has been increased, and if so at what level of significance?   (Cam.)

4. A machine is supposed to produce components to a nominal dimension of 5·000 mm. A random sample of 100 components produced by the machine has a mean of 5·008 mm and a standard deviation of 0·036 mm. Estimate the standard error of the mean and obtain a 95% confidence interval for the mean of the whole output. Test whether, on the evidence of this sample, the mean of the whole output differs from 5·000 mm:

(i) at the 5 % level of significance, (ii) at the 2 % level of significance.
An adjustment is now made to the machine and a sample of 50 components taken at random is found to have a mean of 5·009 mm and a standard deviation of 0·050 mm. Decide whether, on the evidence of this second sample, the machine requires further adjustment. (J.M.B.)

5. A tractor drawbar is intended to have a breaking strain of 12 000 kg. It is known that, because of small variations in the casting, the breaking strain varies between individual bars with standard deviation 400 kg. A sample of 5 bars gave a mean breaking strain of 11 500 kg. Is there significant evidence (at the 5 % level) of a change in breaking strain? Obtain a limit which the new mean breaking strain may be said with 95 % confidence to exceed. (Cam.)

6. A machine is producing components to a nominal dimension of 1·500 units. The dimensions of components in a random sample of eight are found to be 1·502, 1·501, 1·504, 1·498, 1·503, 1·499, 1·505, 1·504. Calculate the standard error of the mean. Assuming the accuracy of this standard error, despite the smallness of the sample, give 95 % confidence limits for the mean component dimension. Is the result consistent with the nominal dimension? (M.E.I.)

7. George has managed a cricket average of about 7 runs per match, with a standard deviation of about 10, for as long as he can remember. His first eight scores this season are 0, 25, 3, 14, 0, 18, 20, 0. Show that the standard deviation is about the same as before. Assuming that the standard deviation can continue to be taken as 10, what score must he make in the 9th match to make his average for these 9 matches just significantly greater than 7? Comment on any assumptions you have to make.

8. An iron foundry asserts that the mean weight of the castings it produces is 20·0 kg. The weights of a sample of 10 castings are: 19·8, 20·3, 20·6, 21·1, 19·3, 19·6, 20·1, 20·8, 21·1, 21·3. Estimate μ and the standard deviation σ of the weight of a casting. Assuming that the true value of σ is equal to your estimated value and that the weight of a casting has a normal distribution, test whether the assertion concerning μ is contradicted significantly at the 5 % level of significance. (Cam.)

9. The table below gives the distribution of age in years at marriage of 175 males:

Age (mid-interval value)	17·5	22·5	27·5	32·5	37·5	42·5	47·5	52·5	57·5	62·5
Frequency	28	68	43	18	9	4	2	1	0	2

Calculate the mean and the standard deviation of these ages. If these ages may be assumed to be a random sample from a large population, calculate 95 % confidence limits for the population mean. Explain briefly whether you would accept or reject the hypothesis that the population mean is 27 years. (M.E.I.)

10. The standard deviation of the number of articles produced in an hour by a factory worker is 14 for all workers, but the mean number produced varies from worker to worker. The numbers produced by a new employee in a representative hour on each of ten successive days were: 108, 124, 92, 113, 129, 146, 117, 103, 131, 128. Do these values show that this worker's rate of production is, at the 5 % level of significance, below the factory average of 127 articles per hour? Determine the two factory average rates from which the average of these ten observations would be judged just to differ at the 5 % level of significance. How many more observations would be needed to reduce to 14 the difference between the two rates defined in this way? (Cam.)

Conclusions from
Sample Data II

13.1. Hypothesis testing: mean from small sample

The supermarket manager has a problem: a customer complains that the weights of her bags of potatoes are consistently below the 2 kg which is stated on the bag to be the average weight of the contents. She says that she has actually weighed the last nine bags and found the mean to be only 1·95 kg. Is the manager to try to persuade her that she was just unlucky, or has she really got a cause for complaint?

In fact, with no more information than is given here, we are unable to decide. We can reasonably assume that the mean weight of 9 bags of potatoes has an approximately normal distribution $N(\mu, \sigma^2/9)$. (Although a sample size of 9 is rather small for a valid application of the Central Limit Theorem, it is likely that the distribution of the weight of a single bag of potatoes will itself be approximately normal and therefore so will be the mean weight of 9 bags.)

It is however impossible to evaluate the claims of rival hypotheses without a knowledge of the size of the population variance σ^2, which the manager is unlikely to have. The best he can do therefore is to obtain the list of weights of the nine bags from the woman, and work out the sample variance s^2 and use this instead.

This is the procedure we adopted in assessing the effect of music in the supermarket in the previous chapter where, with a sample size of 200, we could be fairly confident that the error involved in replacing σ by s would be negligible.

Here, however, if we were to approximate to

$$z = \frac{\bar{x} - \mu}{\sigma/\sqrt{n}} \quad \text{by} \quad \frac{\bar{x} - \mu}{s/\sqrt{n}},$$

the error might be considerable. We will therefore give this second quantity a different name t, so that

$$t = \frac{\bar{x} - \mu}{s/\sqrt{n}}$$

180

If we look at the distribution of t, we will find that it differs from any we have met so far. Because the accuracy of s as an estimate of σ must depend on the sample size n, or more directly on the number of degrees of freedom $v = n - 1$, we find that t has a different distribution curve for each value of v and v is therefore its only parameter. If n is large, the distribution of t will approximate to the distribution of z, which is $N(0, 1)$. We can see this in Fig. 13.1 which shows the distribution of t_v for different values of v and also the $N(0, 1)$ curve for comparison. The percentage points of the t_v-curves are given on page 244, and we can see that, as v increases, the values become very similar to the corresponding normal values and are in fact identical for the line corresponding to $v = \infty$.

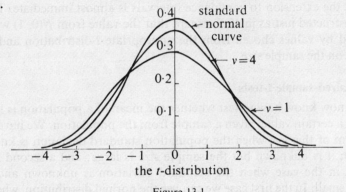

the t-distribution

Figure 13.1

Let us return to the question of the mean weights of the potato bags, and suppose the value of s turns out to be 90 g, that is 0·09 kg. H_0 must be that the weights do satisfy the condition stated, and hence that $\mu = 2\cdot0$ (kg), and the alternative H_1 that $\mu < 2\cdot0$ (kg). The test is one-tailed since the customer is only interested in proving that the bags are underweight.

$H_0 : \mu = 2\cdot0$
$H_1 : \mu < 2\cdot0$
$v = 9 - 1 = 8$

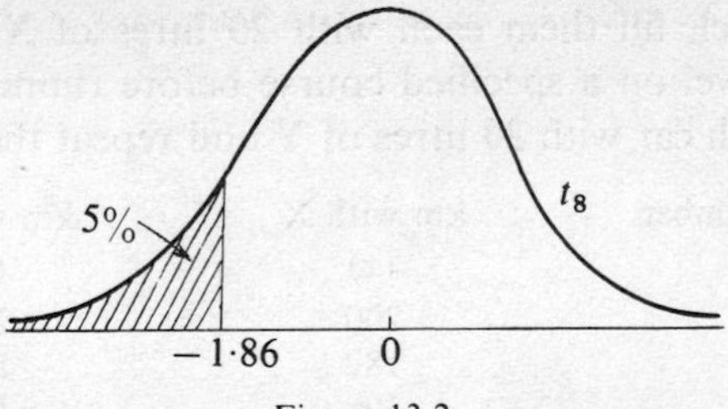

Figure 13.2

We reject H_0 and accept H_1 if $t < k$ where k is the 5% value for t_8. From the t-table the 5% value is 1·86 (remember that we have to look under $P = 10$ as we are carrying out a one-tailed test) and since the t-distribution is, like the normal distribution, symmetric about the origin, the rejection region is $t < -1.86$.

$$\text{But} \quad t = \frac{\bar{x} - \mu}{s/\sqrt{n}} = \frac{1.95 - 2.00}{0.09/\sqrt{9}} = -1.67 > -1.86$$

Hence we must accept H_0 and reject H_1.

We must therefore tell the customer that the evidence is not enough to support her contention. (Notice that if we had used the z-value by mistake instead of the t-value, we would have had the 5% z-value of -1.64 instead of -1.86 and so would have wrongly rejected H_0.)

Now that we have seen how to use the t-distribution for hypothesis testing, the extension to confidence intervals is almost immediate. These are constructed just as in 12.5, except that the value from $N(0, 1)$ will be replaced by values chosen from the appropriate t-distribution and will depend on the sample size.

13.2. Paired-sample t-tests

We now know how to test whether the mean of a population is likely to have a certain value, given a sample from the population. We have met one form of this test when the population standard deviation is known, or when it is unknown but the sample size is large; and a second form for use in the case when the standard deviation is unknown and the sample small. In the first case we can use the normal distribution, whereas in the second we have to use the t-distribution. Suppose now that we are asked to test whether or not two populations have the same mean, given a sample from each of them.

A petrol company marketing brand X is trying to establish that this brand gives a greater number of kilometres per litre than a rival brand Y when tested on a certain popular model of car. They select at random 6 cars of this model, fill them each with 20 litres of X, and record the distance they travel on a specified course before running out of petrol. They then fill each car with 20 litres of Y and repeat the procedure.

car number	km with X	km with Y
	(x)	(y)
1	200	208
2	182	174
3	198	194
4	174	173
5	191	186
6	213	205

Our null hypothesis H_0 must be that there is no difference between X and Y and hence that the mean number μ_1 of kilometres travelled on 20 litres of X by any car of this model is equal to the mean μ_2 travelled on 20 litres of Y. Therefore H_0 is that $\mu_1 = \mu_2$. As the petrol company is concerned to prove the superiority of their brand, they will consider only the alternative H_1, that $\mu_1 > \mu_2$, and hence the test is a one-tailed test only. (If the test were being carried out by a neutral body, they would have had to consider the two alternatives $\mu_1 > \mu_2$ and $\mu_1 < \mu_2$, and the test would then have been two-tailed.)

Now we have two samples to consider, each with its own mean and standard deviation, and we want to establish a method for deciding whether or not the difference in the sample means can be attributed purely to sampling.

This seems, on the face of it, to be a much more complex problem than the previous ones we have had with a single sample, so it would seem reasonable to try to reduce it to a single sample problem. If we think about it, we are not in this case really concerned about the performance of each car with the two brands of petrol separately but only with the difference in performance. For instance with car 1 it is not really the 200 and 208 kilometres which matter in themselves, but only the 8 kilometres difference. It would seem a logical step to dispense with the actual distances and simply consider their difference $d = x - y$.

car number	difference in km (d)
1	-8
2	8
3	4
4	1
5	5
6	8

What would the expected value of the difference be if the expected value of x is μ_1 and that of y is μ_2? We saw in Chapter 11 that

$$E(d) = E(x - y) = E(x) - E(y) = \mu_1 - \mu_2.$$

So, if H_0 is true we have $E(d) = 0$. If H_1 is true we have $E(d) > 0$. We do not know the standard deviation of the population of differences, but we can find an estimate of it from the sample. Hence, if we are prepared to assume that the distribution of the mean difference is at least approximately normal, we are now faced with a problem we can solve, that of testing whether the mean of a population has a given value. (We would really have to test the normality assumption as our sample size is only 6, but it seems a reasonable one.) Of course since our sample is small, we will need to use the t-distribution.

$H_0 : \mu_1 = \mu_2$, i.e. $E(d) = 0$.
$H_1 : \mu_1 > \mu_2$, i.e. $E(d) > 0$.
$v = 6 - 1 = 5$. Hence from the t-tables we reject H_0 at 5% significance level if $t > 2 \cdot 02$.
But $\bar{d} = (-8 + 8 + 4 + 1 + 5 + 8)/6 = 3$.
$s^2 = \{(8^2 + 8^2 + 4^2 + 1^2 + 5^2 + 8^2) - 6 \times 3^2\}/5 = 36$
$$t = \frac{3 - 0}{6/\sqrt{6}} = 1 \cdot 22.$$

Hence we accept H_0.

Therefore the company would be forced to conclude that the test did not prove that brand X was superior.

Notice that this example was a rather special one because the members of the two samples came in pairs, each pair corresponding to the two results for one car. For this reason the test which was used is called a *paired-sample t-test*. This test is normally used for samples naturally paired in some way.

13.3. Two sample t-tests

We must now consider what happens if we have two samples which do not "pair off" naturally, but which are independent.

Bunbury Grammar School has recently installed a language laboratory. The headmaster wishes to find out whether the children learn French more or less successfully if they have lessons in the laboratory rather than ordinary teaching. He takes two *similar* classes (chosen to have as far as possible an equal spread of ability in each class), A and B, and lets them learn French for a year, class A in the laboratory and class B in the ordinary way. At the end of the year he gives both classes the same test.

The headmaster wishes to compare the average marks obtained by the two classes, when these are thought of as a sample of the children being taught by the two methods; so his null hypothesis must be that the mean population marks are the same, $\mu_1 = \mu_2$. (Strictly speaking he may only be comparing the average mark obtained by the teaching of one particular teacher with the language laboratory, and he would be better to enlarge his experiment to cover this.) For the results he has, his alternative hypothesis must be that $\mu_1 \neq \mu_2$ since he is interested in knowing whether the laboratory will increase or decrease the marks.

Suppose there are 15 children in class A and their average mark in the test $\bar{x}$ is 60%, and that there are 12 children in class B and their average $\bar{y}$ is 50%.

We are probably justified in assuming that $\bar{x}$ is approximately $N(\mu_1, \sigma_1^2/15)$ and $\bar{y}$ is $N(\mu_2, \sigma_2^2/12)$, where σ_1^2 and σ_2^2 are the unknown variances of the two populations of marks. Again we are not really concerned with $\bar{x}$ and $\bar{y}$ separately, but only with the difference between the groups; and so it seems reasonable to consider $(\bar{x}-\bar{y})$ which we would expect to be distributed as $N(\mu_1-\mu_2, \sigma_1^2/15+\sigma_2^2/12)$.

If H_0 is true we have $\mu_1 = \mu_2$, and hence $E(\bar{x}-\bar{y}) = 0$. Provided we know σ_1 and σ_2, we are faced with the problem of testing whether a statistic has zero mean, which we saw how to solve in Chapter 12. In the more usual case (and certainly in our present example) σ_1 and σ_2 will not be known; we will have to estimate them. The most obvious solution would be to substitute separate estimates for σ_1 and σ_2 formed from the appropriate samples. This procedure is fine provided the sample sizes are fairly large. For smaller samples, such as we have here, we shall have to make the assumption that $\sigma_1 = \sigma_2$. In general this will be a reasonable assumption to make in the sort of populations we shall meet (strictly speaking this assumption should be first tested, but the test is beyond the scope of this book). Making this assumption here, we have $\bar{x} - \bar{y}$ distributed as $N(\mu_1 - \mu_2, \sigma^2(\frac{1}{15} + \frac{1}{12}))$ where σ is the common value of σ_1 and σ_2.

Now we are in the situation of 13.1. We must substitute a sample estimate s for σ and then use the statistic

$$t = \frac{(\bar{x}-\bar{y})-(\mu_1-\mu_2)}{s\sqrt{\{\frac{1}{15}+\frac{1}{12}\}}}$$

which will have the t-distribution. But what should our estimate s be? We now have not one sample but two from which to estimate σ; we should use all the information we have, and take both samples into account. We cannot apply the standard formula and simply sum over both samples, as each will have a different sample mean. So we will have to take the sum of squares of deviations about its own mean for each sample separately, add the resulting values and divide by the "appropriate" number of degrees of freedom in order to make the estimate unbiased. What will this "appropriate" number be? We have

$$E(\Sigma(x-\bar{x})^2) = (15-1)\sigma^2, \; E(\Sigma(y-\bar{y})^2) = (12-1)\sigma^2$$

so that $$E(\Sigma(x-\bar{x})^2 + \Sigma(y-\bar{y})^2) = 25\sigma^2.$$

So our estimate of σ^2 should be $\frac{1}{25}[\Sigma(x-\bar{x})^2+\Sigma(y-\bar{y})^2]$ and our t-distribution will have 25 degrees of freedom.

In the more general case in which the first sample contains m elements and the second sample n elements, the corresponding value of s^2 will be

$$\frac{\Sigma(x-\bar{x})^2 + \Sigma(y-\bar{y})^2}{m+n-2}$$

and the number of degrees of freedom for the t-distribution, which is always equal to the denominator of the variance estimate, will be $m+n-2$.

We can now carry out the test exactly as in 13.1. Suppose we have $\Sigma(x-\bar{x})^2 = 1100$ and $\Sigma(y-\bar{y})^2 = 900$, then $s^2 = (1100+900)/25 = 80$.

$H_0 : \mu_1 = \mu_2$

$H_1 : \mu_1 \neq \mu_2, \qquad v = m+n-2 = 25.$

Table 3 does not give us values for $v = 25$. It does tell us that the 5 % value for $v = 24$ is 2·06 and for $v = 30$ is 2·04. Hence the value we require is between these two.

$$t = \frac{10-0}{\sqrt{[80(\frac{1}{15}+\frac{1}{12})]}} = 2\cdot88, \text{ hence reject } H_0.$$

So we must conclude that there is a significant difference in the marks.

13.4. Testing differences in proportions

Suppose the difference we are considering concerns differences in proportions rather than means. For example:

"I can't think why we ever gave women the vote," said George, "they don't know a thing about politics!"

"They're no worse than men," I replied, "I bet if you asked people the name of the Foreign Secretary, the men would be no more likely to know than the women."

"I'll take you up on that," said George, "We'll test it."

George went out and asked 30 men and 12 answered correctly. When he learned that out of the 20 women I had asked only 4 had given the right answer, he was very smug. However I insisted upon a proper statistical comparison.

If we were only considering the number of men who answered correctly, x_m, then we would expect x_m to be distributed $B(30, p_m)$ where p_m is the probability that any man answers correctly. However we cannot compare x_m and x_w (the number of women who answer correctly) directly, since the sample sizes are different. It is therefore the proportions $x_m/30$ and $x_w/20$ which are important.

Since the sample sizes are quite large we have x_m approximately distributed as $N(30p_m, 30p_m q_m)$ and so we have $x_m/30$ is approximately $N(p_m, p_m q_m/30)$. Similarly $x_w/20$ is very approximately $N(p_w, p_w q_w/20)$.

What about our hypotheses? The null hypothesis must be that the probabilities of answering correctly are equal so that $p_m = p_w$. H_1 will be $p_m > p_w$ since George is only concerned to prove male superiority, and the test is therefore one-tailed.

Since again we are only concerned with the difference in proportions, we should look at $\frac{1}{30}x_m - \frac{1}{20}x_w$ which will be distributed as

$N(p_m - p_w, \frac{1}{30}p_m q_m + \frac{1}{20}p_w q_w)$, provided the two samples are independent.

We obviously need to know the values of p_m and p_w to determine the variance, but as these are the population parameters in which we are interested, we will have to estimate them from the samples. For the purposes of the test, we assume that H_0 is true, which means that $p_m = p_w$; we can therefore treat both samples as a single sample with 50 members and estimate the common value, p say, of the proportion answering correctly. Since altogether 16 people answered the question correctly, our best estimate is $\frac{16}{50} = 0.32$.

So $\frac{1}{30}x_m - \frac{1}{20}x_w$ is approximately $N(0, 0.32 \times 0.68 \times (\frac{1}{20} + \frac{1}{30}))$ provided H_0 is true.

We can now apply the test.

$H_0 : p_m - p_w = 0$

$H_1 : p_m - p_w > 0$. Reject H_0 at 5% if $z > 1.64$.

But $z = (\frac{12}{30} - \frac{4}{20})/\sqrt{\{0.32 \times 0.68(\frac{1}{20} + \frac{1}{30})\}} = 1.49$.

Hence H_0 cannot be rejected.

In general this test can be used for samples of sizes n_1 and n_2 if $n_1\hat{p}$, $n_2\hat{p}$, $n_1(1 - \hat{p})$ and $n_2(1 - \hat{p})$ are all greater than 5, where $\hat{p} = \dfrac{x_1 + x_2}{n_1 + n_2}$. For greater accuracy a form of 0.5 correction can be applied but this usually only affects just significant results from fairly small samples.

EXERCISE 13A

1. A sample of 10 observations is taken from a normal distribution which has assumed mean 2·0 and unknown variance. The sample has $\bar{x} = 1.4$ and $s^2 = 2.7$. Is the assumed mean correct?

2. In a small factory manufacturing large pink teddy bears, with a mean height of 0·99m, a new woman is taken on to do the cutting out. It is found that the first 14 teddy bears she makes have an average height of 0·89 m with a standard deviation of 0·154 m. Is this significant evidence that the new woman cuts out smaller teddy bears?

3. The following frequencies of the number (r) of disintegrations in a given time were obtained in 30 observations of a Wilson cloud chamber.

Number of disintegrations (r)	0	1	2	3	4	5
Frequency (f_r)	5	8	4	6	4	3

   Plot these frequencies in a suitable diagram. Calculate estimates of the mean and standard deviation of r. Obtain limits which you would expect with 95% confidence to enclose the population mean of r. (Cam.)

4. Fifteen observations were taken from a population as follows:

   1·41, 1·41, 0·23, 0·36, −0·01, 0·07, −0·10, 2·20, −0·62, −0·66, 0·30, 0·07, −0·16, 0·02, 0·66.

   Test the hypothesis that the mean of the population is zero and give 95% confidence limits for the mean. Describe the meaning of the confidence limits. (M.E.I.)

5. The following determinations of the speed of sound in air were made in independent repetitions of the same experimental procedure (the unit being 10^4 cm/s).

$$3·35, \quad 3·27, \quad 3·30, \quad 3·33, \quad 3·32, \quad 3·29.$$

Obtain an estimate based on all the determinations. Also estimate σ, the standard deviation of a single determination. Find an estimate of the standard deviation of your estimate of the speed of sound, and give 95% confidence limits for the speed of sound. (Cam.)

6. A sample of six independent observations of a normal variable x gives $\Sigma x = 18·84$, $\Sigma x^2 = 60·66$. Use this sample to obtain a 95% confidence interval for the mean of the variable x. Another sample of ten independent observations gives $\Sigma x = 31·50$, $\Sigma x^2 = 102·83$. Calculate the best 95% confidence interval for the mean that you can, and explain in what sense it is "best". (Cam.)

7. Nine pairs of men were chosen so that initially each pair were of the same weights and were of the same age and income groups. A special drug was administered to one of each pair, and the weight gains in kg over three months were as given below.

Pair	1	2	3	4	5	6	7	8	9
With drug	3	4	3	1	−1	2	3	3	2
Without drug	2	1	−4	0	−1	3	−1	0	2

Is there any evidence that the drug produced a greater gain in weight than would normally be expected?

8. Two different methods for the determination of a certain additive in steel were compared by taking nine different steels. A strip of each steel was cut in two and the two halves were analysed by the two different methods, allocated at random. Use the following results, given in coded units, to compare the two methods of analysis.

Steel strips	1	2	3	4	5	6	7	8	9
Method A	21·2	23·9	0·4	2·5	0·3	7·6	4·2	2·9	1·4
Method B	20·8	26·9	0·3	0·1	0	7·6	2·7	4·6	2·7

(M.E.I.)

9. In order to test the effect of a rust-proofing process on the strength of metal, 100 pieces are cut in two; one of each pair is treated and the other left untreated. In suitable units the average value of the excess of the strength of the treated piece over that of the corresponding untreated piece is 2·1 and the standard deviation is 10. Explaining carefully what assumptions are made and what the conclusion means, carry out a significance test at the 5% level to discover whether the treatment has any effect on the strength. (J.M.B.)

10. An experiment was performed to compare two methods of analysing the percentage impurity in a certain chemical. Tests were made on twelve different samples of the chemical, each sample being analysed by both methods. The results of the experiment are given in the following table:

Sample No.	1	2	3	4	5	6	7	8	9	10	11	12
Method A	2·12	2·45	2·43	2·51	2·42	2·44	2·56	2·41	2·41	2·46	2·43	2·38
Method B	2·26	2·45	2·46	2·44	2·55	2·56	2·65	2·36	2·50	2·52	2·48	2·42

Find the standard error of the mean of the differences in the results for these samples. Hence test, at both the 5% and 1% levels of significance, the hypothesis that on average the two methods of analysis are equivalent.

(M.E.I.)

11. In order to find out whether the average speed of motor vehicles leaving London is different from that of those entering London, cars and motorcycles were timed over a stretch of the Portsmouth road.

	Leaving London	Entering London
No. of vehicles timed	50	50
Mean time in seconds	17·04	18·38
Variance in seconds$^2 = S^2$	8·846	9·106

Determine whether the difference between the means is significant at the:
(i) 5 %, (ii) 1 % level.
Comment on the meaning of the results.   (J.M.B.)

12. The heights (in cm) of small random samples of boys and girls of age 13 were:

Boy	150	145	148	155	160	143	140
Girl	143	148	130	125	148	150	143	140

Carry out a 2-sample t-test to test whether the mean heights of the boys and girls differ at the 5 % level of significance. If you were asked to test whether the mean height of the boys differed from 150 cm, would you use a standard error estimate based only on the boys' heights or, as in the test just performed, one based on the heights of all the children? Justify your answer.   (Cam.)

13. A sample of 500 long-playing records made by company A has a mean playing time of 26·4 minutes and the standard deviation S of the playing times is 5 minutes, while for a sample of 320 records made by company B the mean and standard deviation are 27·3 and 4 minutes respectively. Determine whether the mean playing times differ significantly between records made by the two companies.   (J.M.B.)

14. One group of mice was fed a normal diet and another group was fed a test diet. Examine the following data for differences in the average weight in the two groups.

Body weight (g)

Normal diet	40	30	41	41	41	42	31
Test diet	34	28	41	38	34	41	35

(M.E.I.)

15. The following table is taken from a recent report on the serving of school dinners in grammar schools of Great Britain:

Time taken for school dinner (minutes)	20	25	30	35	40	Total
No. of girls' schools	5	19	125	65	49	263
No. of co-educational schools	10	35	121	56	44	266

Calculate the mean and standard deviation of the times taken:
(i) in girls' schools,
(ii) in co-educational schools.
Determine whether or not the difference between the means is significant.
(J.M.B.)

16. A machine is claimed to produce brass pins of average length 3·50 units. A random sample of twelve pins is chosen and the length x of each pin is measured. It is found that $\bar{x} = 3·513$, $\Sigma(x - \bar{x})^2 = 0·00372$. Assuming that the length of the pins produced is normally distributed, test the hypothesis that the claim made for the machine is true. A second random sample of 6 pins is taken a week after the first sample. For this sample $\bar{x} = 3·502$ and $\Sigma(x - \bar{x})^2 = 0·00215$. Test whether the performance of the machine has improved.
(Cam.)

17. Comment on the following quotations:
    (a) "The fact that there is no significant difference between the means proves that the two groups can be regarded as equivalent."
    (b) "The sample means differed at the 5% level of significance which shows that the probability that the true means are the same is less than 1 in 20."
    (Cam.)
18. The body-work on two well-known models was compared on cars that were four years old. Of 20 cars of the first model 9 showed a considerable degree of rusting, whereas only 6 out of 25 of the second model did so. Is there any evidence that the finish of the second model is superior to that of the first?
19. An experiment was performed to investigate the toxicity of a possible flavouring agent in a domestic product. 208 mice were fed on the product without the agent and 53 developed neoplasms, whereas 32 out of 104 mice developed neoplasms when 1% of the agent was added to the product. Compare the difference in proportions of mice showing the symptoms and state whether or not the flavouring agent may be regarded as safe.   (M.E.I.)
20. Four-fifths of the families in one street owned a car whereas of exactly the same number of families in a neighbouring street three-fifths did. How many families would there need to be in each street for the difference to be significant at the 5% level.

13.5. The sign test

In the tests we have met so far concerning the "average" of a population we have formulated our hypothesis about the mean of the population and have had to assume either that the distribution of the population was normal, or that the sample size was so large that the Central Limit Theorem could be applied.

We will now consider some *non-parametric* (or *distribution-free*) tests which make no assumptions about the population distribution. Such tests are usually easier to make about the median m of the distribution rather than the mean.

For example, suppose it was suspected that twins entering primary school tend to have a lower verbal ability than expected due to the fact that up to then their communication has largely been with each other. In order to test this, one child out of each of the sixteen pairs of twins of this age in a particular educational district is given a test of verbal ability on which the standard score for this age is 100. The results are:

86	95	73	88	112	103	100	79
103	93	108	70	85	94	122	82

Assuming that the test score is standardized to give a symmetric distribution, the median score m for the total population of school children will also be 100. We should then have, for the variable x, representing any member of the total population, except those equal to 100, $P(x < 100) = P(x > 100) = \frac{1}{2}$.

We can use this fact to test the null hypothesis H_0 that the scores of the twins come from the population with median m equal to 100, against the alternative H_1 that they come from a population with $m < 100$.

We need to consider whether each score is greater, equal to, or less than 100, and can hence replace the actual scores by $+1$, 0, or -1 respectively, as follows:

$$
\begin{array}{cccccccc}
-1 & -1 & -1 & -1 & +1 & +1 & 0 & -1 \\
+1 & -1 & +1 & -1 & -1 & -1 & +1 & -1
\end{array}
$$

If we now count only non-zero values, out of a total of 15 results, 5 are equal to $+1$. But assuming H_0, that $m = 100$, each value should have an equal probability of being $+1$ or -1, and the number of values r equal to $+1$ should therefore have the distribution $B(15, \frac{1}{2})$.

H_1 indicates that under the alternative hypothesis we would expect r to be small. From the table on page 168 we find that $P(r \leqslant 3) = 0 \cdot 017$, and $P(r \leqslant 4) = 0 \cdot 059$ for the distribution $B(15, \frac{1}{2})$. Hence the rejection region at the 5% significance level is $r \leqslant 3$ for a one-tailed test.

Here we have $r = 5$, and must therefore accept that the results do not give sufficient evidence that twins have lower verbal ability scores than other children at this age.

We can thus test whether any sample comes from a population with given median m_0 by replacing each reading by $+1$, 0, or -1, according as it is greater, equal to, or less than m_0 and counting the number of values r equal to $+1$ and the total number n of non-zero values. Only if r lies outside the acceptance region for $B(n, \frac{1}{2})$ will we reject the null hypothesis that the sample comes from a population with median m_0. If the sample size is large, we can use the normal approximation to $B(n, \frac{1}{2})$.

The sign test is less powerful than the t-test since it does not make use of all the available information (for instance, in the above example both 122 and 103 were assigned the values $+1$). However its strength lies in that it is more widely applicable. Suppose the twins had been of different ages and we had been told only that each child's score was above or below the standard score for his or her particular age. We could still have applied the sign test as we would be able to assign $+1$, 0, or -1 to each score.

The sign test can easily be extended to test whether two samples of paired values come from the same distribution, and thus can be used as one non-parametric equivalent of the paired-sample t-test. The test is carried out by calculating the differences between each of the pairs and then applying a sign test with $m_0 = 0$ to the differences. If the parent distributions were identical, we would expect the differences to come from a distribution with median zero.

Suppose we had tested both children in each of the sets of twins which contained a boy and a girl, in order to see whether there was any evidence that either sex was superior on verbal ability scores.

Girl's score	95	112	103	92	79	122
Boy's score	87	93	104	82	82	115
Difference	+8	+19	−1	+10	−3	+7
Assigned value	+1	+1	−1	+1	−1	+1

H_0: median of differences is zero and hence the number r of $+1$ values
has the distribution $B(6, \frac{1}{2})$

H_1: median of differences is not zero (2-tailed test)

Given H_0, $P(r \leqslant 0 \text{ or } r \geqslant 6) = 2(\frac{1}{2})^6 \simeq 0.031$
$$P(r \leqslant 1 \text{ or } r \geqslant 5) = 2[6(\frac{1}{2})^6 + (\frac{1}{2})^6] \simeq 0.219$$

Hence the rejection region for a 5% level test (2-tailed) contains only $r = 0$ and $r = 6$. We therefore accept H_0 and conclude that the results do not show that either sex has superior verbal ability in this age-group.

13.6. The Wilcoxon signed rank test of differences

In the comparison between scores of girls and their twin brothers, the sign test did not take into account that in the only two cases where the boys' scores were superior, the differences were marginal, whereas in the four other cases the girls had much higher scores than the boys.

There is a test which is a compromise between the sign test and the t-test, being more precise than the former but not, like the latter, requiring any assumptions about the underlying distribution of the scores. The test is known as the *Wilcoxon signed rank test of differences* and is a non-parametric test applicable in paired sample cases.

It takes into account not the actual differences but their relative size, by *ranking* the non-zero differences. In order to rank the differences we arrange them in ascending order of *absolute* size and assign a rank 1 to the smallest, 2 to the next smallest, and so on. (Should it happen that two or more values are equal, they would be given the same rank, dividing up equally the sum of the ranks they would have obtained in the normal way, e.g. if the 4th and 5th were equal, they would each receive rank 4·5; if the 4th, 5th and 6th were equal, they would each be given rank 5. Such ranks are said to be "tied.")

Using the same example as above, we have:

Difference between scores
of boy and girl twin $+8$ $+19$ -1 $+10$ -3 $+7$

Rank of absolute difference   4   6   1   5   2   3

Rearranging these in rank order:

Rank of difference   1   2   3   4   5   6
Sign of difference $-$ $-$ $+$ $+$ $+$ $+$

(The negative signs occur where the boy's score was higher.)

Whichever is the smaller, the sum of the ranks which have a positive sign or the sum of those with a negative sign, is called T. Here it is the negative sum which is smaller and so $T = 1 + 2 = 3$.

We can then decide how many ways there are of obtaining a value of T as small as this, by assigning a positive or negative sign to each rank as we please.

We can tabulate the possibilities systematically as follows:

Ranks	1	2	3	4	5	6	
	$+$	$+$	$+$	$+$	$+$	$+$	$T = 0$
	$-$	$+$	$+$	$+$	$+$	$+$	$T = 1$
	$+$	$-$	$+$	$+$	$+$	$+$	$T = 2$
	$+$	$+$	$-$	$+$	$+$	$+$	$T = 3$
	$-$	$-$	$+$	$+$	$+$	$+$	$T = 3$

Any other way of assigning signs gives a value of T which is greater than 3, except those which are obtained from those shown above by interchanging $+$ and $-$ signs. Since we were not initially prejudiced about whether boys or girls would score higher, the test is two-tailed, and we will therefore take into account these other small values of T where T is the sum of the positive ranks. If the test were one-tailed, we would only count the 5 shown above.

Since each rank can be given a sign independently of all the others, the total number of ways of assigning the signs is 2^6. But if the null hypothesis holds, both girls' and boys' scores come from the same population, and hence all of these arrangements are equally likely.

Given H_0, $P(T \leq 0) = \dfrac{2}{2^6} \simeq 0.031$, $P(T \leq 1) = \dfrac{2 \times 2}{2^6} \simeq 0.063$. Hence the rejection region for a 5% significance level only includes $T = 0$.

As in the sign test, we therefore accept the null hypothesis.

We can see that if the number of pairs n is large, this method will become very tedious. However, luckily, for $n \geqslant 25$ or so, the distribution of T becomes approximately normal and hence we can use the standard normal distribution to work out the limits of the acceptance region.

We can work out the mean and standard deviation for this normal distribution if we think of T as being a sum which will include each of the numbers r between 1 and n, each with probability $\frac{1}{2}$, depending on the nature of the sign which accompanies the rank r.

We can therefore write: $T = x_1 + x_2 + \ldots + x_n$, where x_r is an independent random variable such that $P(x_r = 0) = \frac{1}{2}$, and $P(x_r = r) = \frac{1}{2}$, for $1 \leqslant r \leqslant n$. We have $E(x_r) = \frac{1}{2}r$, $\text{Var}(x_r) = E(x_r^2) - (\frac{1}{2}r)^2 = \frac{1}{4}r^2$.

Then $E(T) = E(x_1) + E(x_2) + \ldots + E(x_n) = \frac{1}{2}\overset{n}{\underset{1}{\Sigma}}r = \frac{1}{4}n(n+1)$.

$\text{Var}(T) = \text{Var}(x_1) + \text{Var}(x_2) + \ldots + \text{Var}(x_n) = \frac{1}{4}\overset{n}{\underset{1}{\Sigma}}r^2$
$= \frac{1}{24}n(n+1)(2n+1)$.

Hence, for $n \geqslant 25$ or so we can use $N(\mu, \sigma^2)$ as an approximation for the distribution of T, where $\mu = \frac{1}{4}n(n+1)$ and $\sigma^2 = \frac{1}{24}n(n+1)(2n+1)$.

13.7. Non-parametric tests for two unpaired samples

We can also use a "ranking" method when we wish to decide whether two unpaired, and possibly unequal, samples may have come from the same population.

Suppose we wish to investigate whether British shipyards are really less profitable than those in other European countries. Exact figures of the profits over the last three years might be difficult to obtain, and might not be strictly comparable due to different factors operating in different countries, e.g. government grants, tax, and so on.

It might, however, be possible to select four British and five comparable European shipyards and put them in order of their recent profit figures. The result might be:

Rank	1	2	3	4	5	6	7	8	9
Country	B	B	E	B	B	E	E	E	E

where British and European yards are indicated by B and E respectively, and the yards are ranked in ascending order of profitability.

In this case, of course, we could not use a parametric test like the t-test. However, even if the exact figures were known, we could still arrive at a rank order like this by putting the samples together and ranking them as if they were all in one sample, remembering to allow for "tied" ranks.

Here our null hypothesis is that there is no real difference in profitability between British and European shipyards; the alternative is that European yards are more successful.

We proceed as in the Wilcoxon signed rank test, and calculate that $R_B = 1 + 2 + 4 + 5 = 12$, where R_B is the sum of the ranks of British yards.

We can again use a trial-and-error approach to see what the probability is of getting an even smaller value for R_B. However there is a slight difference in that here the total number of both Bs and Es is fixed at 4 and 5 respectively, so the letters assigned to each rank are not independent as the signs were in the previous example.

We have, working systematically up from the smallest value of R_B:

Rank	1	2	3	4	5	6	7	8	9	
Country	B	B	B	B	E	E	E	E	E	$R_B = 10$
	B	B	B	E	B	E	E	E	E	$R_B = 11$
	B	B	B	E	E	B	E	E	E	$R_B = 12$
	B	B	E	B	B	E	E	E	E	$R_B = 12$

All other arrangements give $R_B > 12$, and we do not consider the ones which give small values of R_E since the test is one-tailed.

There are in all $\binom{9}{5} = 126$ possible arrangements of the letters, all equally likely if we assume the null hypothesis holds, so $P(R_B \leqslant 12)$ $= \frac{4}{126} \simeq 0.032 < 0.05$. Hence our value $R_B = 12$ is in the rejection region for H_0, and we therefore agree to accept the alternative that British shipyards are for some reason less profitable than European ones.

The method becomes more tedious as the number n of ranks (with both samples combined) increases but, as before, the distribution of R can be approximated by a normal distribution provided the number in each sample is greater than eight. If there are n_1 in the first sample and n_2 in the second, then the total sum of all the ranks will be $\frac{1}{2}n(n+1)$ $= \frac{1}{2}(n_1 + n_2)(n_1 + n_2 + 1)$. If the null hypothesis holds we would expect the sum to be divided proportionately between the samples, giving:

$$E(R_1) = \frac{n_1}{n} \times \frac{1}{2}n(n+1) = \frac{1}{2}n_1(n_1 + n_2 + 1).$$

Although more difficult, it can be proved that

$$\text{Var}(R_1) = \frac{1}{12}n_1 n_2 (n_1 + n_2 + 1).$$

We can therefore use the normal distribution with this mean and variance to find the limits of the acceptance region for R_1 for $n_1, n_2 \geqslant 8$.

To get an idea of how accurate the normal distribution is as an approximation, we could try using it on the shipyard figures, although as here $n_B = 4$ and $n_E = 5$, we would not expect a very good fit.

We have $\qquad E(R_B) = \frac{1}{2}n_B(n_B + n_E + 1) = 20$

$$\text{Var}(R_B) = \frac{1}{12}n_B n_E(n_B + n_E + 1) = \frac{50}{3}$$

$$R_B = 12, \text{ so } z = \frac{12 - 20}{\sqrt{(50/3)}} = -1.96$$

Since $z < -1.64$, the 5% limit for a one-tail test, reject H_0. In fact

$P(z \leqslant -1.96) \simeq 0.025$, which is not too different from $P(R_B \leqslant 12)$ $\simeq 0.032$ which was the figure obtained directly. Even for these small samples then, the corresponding normal distribution gives a reasonable approximation.

When $n_1 = n_2$, the test is known as the *Wilcoxon composite ranks test*. Whether or not the sample sizes are equal, it gives results identical to a test called the Mann-Whitney U test, although the procedure for arriving at the result is often quoted in a slightly different form. The distributions of the statistic in this and the previous section have been evaluated assuming that there are no tied ranks. When there are ties, some modifications have to be made, but these will not be important provided the proportion of ties is small.

EXERCISE 13B

1. The length of time 7 employees stay with a certain company are given below in years. Test the hypothesis that the median length of stay is 5 years against the alternative that it is not equal to 5 years.

    15·0   0·6   2·2   4·3   7·4   1·1   4·6

2. George's cricket scores for the first nine matches of the season were 8, 25, 8, 14, 10, 18, 20, 7, 0. Can you accept the hypothesis that his median score is 7, even though he asserts that it is more than 7?

3. Use the sign test to test the difference between the distances travelled with petrol X and Y for the data of §13.2.

4. Eight volunteers are asked to perform a test of manual dexterity before and after they have consumed two pints of beer. Do the results given below support the hypothesis that alcoholic consumption reduces the level of performance?

Volunteer	1	2	3	4	5	6	7	8
1st test	32	54	22	43	40	37	16	48
2nd test	21	49	26	15	43	28	17	30

5. 25 girls and 30 boys are given an intelligence test. When their scores are ranked, the sum of the girls' ranks is 775. Test the hypothesis that boys and girls are equally intelligent.

6. Two samples have $n_1 = 2$ and $n_2 = 5$. Assuming that none of the values are equal, find the probability distribution of R_1 and hence its mean and variance.

7. The 12 competitors in Bunbury School 100 metres race in 1975 had times in seconds: 11·6, 12·4, 14·4, 13·2, 12·8, 14·6, 11·8, 13·6, 12·1, 14·5, 13·1, 14·0. In 1970, 11 competitors had times: 12·9, 14·3, 11·4, 12·6, 13·4, 12·6, 11·5, 12·9, 12·0, 12·6, 12·3. Is there any evidence that the second set of times is faster than the first?

The data from the following questions can also be used for non-parametric tests (replacing "mean" by "median" where necessary): Ex. 12B, nos. 6–10; Ex. 13A, nos. 1, 4, 7, 8, 10, 12, 14.

Testing Goodness of Fit and Association

14.1. Probability models

In some of the previous chapters we have discussed the building of "probability models", where we take a real situation and, on the basis of our knowledge of that situation (which will almost certainly include some experimental data) try to fit a theoretical distribution to it. We are particularly pleased if it looks as though it will be well fitted by one of the standard distributions (uniform, binomial, geometric or Poisson if discrete, and uniform, exponential or normal if continuous) since we can then readily use a number of results which are standard for these distributions.

Naturally no real-life data fit a distribution exactly; if they did, we should suspect that the results had been "fiddled". The question arises as to how much discrepancy we ought reasonably to allow. At what point ought we to draw a line and say that the data differ too much from the results predicted by the theoretical distribution?

14.2. Waiting times

Suppose we follow up the situation in Chapter 9 where we were interested in the time spent waiting in a certain bus queue. In order to specify our model we assumed that the buses ran at 10-minute intervals

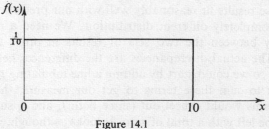

Figure 14.1

197

and that people joined the queue at random times. This led us to the first model, which was that the time spent waiting for the bus was uniformly distributed in the interval 0–10 (Fig. 14.1).

Any data we collect to test the model will necessarily have to be discrete. It is likely that waiting times will be measured to the nearest minute—and so we will only be able to test whether the model and the data agree reasonably well in the numbers of waiting times lying in the intervals 0–0·5, 0·5–1·5, 1·5–2·5, . . ., 9·5–10. Suppose we take 200 waiting times, then the model would predict that $\frac{1}{20}$ of them, that is 10, would be between 0 and 0·5 minute, since this interval is $\frac{1}{20}$ of the total length. Similarly the model predicts $(\frac{1}{10}) \times 200 = 20$ would lie between 0·5 and 1·5 minutes, and so on. We can draw up an expected histogram of the results.

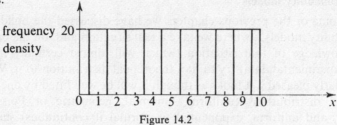

Figure 14.2

(Notice that the two end intervals are half as long as the others.)

Suppose we now observed the queue, selected a random sample of 200 people, and recorded the length of time they waited for a bus with the following results. (In this case no-one waited more than 10 minutes.)

x	0	1	2	3	4	5	6	7	8	9	10
f	24	13	22	17	19	11	18	24	21	25	6

We can now draw up a table comparing the expected and observed frequencies.

x	0	1	2	3	4	5	6	7	8	9	10
f_o	24	13	22	17	19	11	18	24	21	25	6
f_e	10	20	20	20	20	20	20	20	20	20	10

f_o is the observed frequency and f_e the expected frequency.

Do these results fit reasonably well with our prediction, or must we choose a completely different distribution? We need a measure of the discrepancy between the two sets of results in order to answer this question. The actual discrepancies are the differences between f_o and f_e for each x, so we could start by adding a line tabulating $f_o - f_e$. However if we were to sum these terms to get our measure, the positive and negative terms would cancel out (since both f_o and f_e sum to 200) and we would be left with a total of zero. It looks as though, just as with the standard deviation, it would be more appropriate to work with the squared discrepancies. So we add a line for $(f_o - f_e)^2$.

Do we simply want to form the sum of this line to get our measure? Notice that for both $x = 10$ and $x = 7$ we have a difference of 4 between the expected and observed frequencies. However for $x = 10$ the difference is between 6 and 10 and is 40% of the expected frequency, whereas for $x = 7$ the difference is between 20 and 24 and is only 20%. If we let these both make the same contribution to the total sum, we would be tending to underestimate the difference for the case $x = 10$ which is proportionately greater. It might seem, therefore, that we should work with the difference as a proportion of the expected frequency and use the sum of the terms

$$\left(\frac{f_o - f_e}{f_e} \right)^2.$$

Now, however, we are saying that the difference between an observed frequency of 6 and an expected one of 10 is the same as the difference between an observed frequency of 60 and an expected one of 100, or an observed frequency of 600 and an expected one of 1000, whereas it seems that the last should be the most important. So we will compromise between basing the measure on actual differences and on proportions and use $(f_o - f_e)^2 / f_e$. The table now becomes:

x	f_o	f_e	$f_o - f_e$	$(f_o - f_e)^2$	$(f_o - f_e)^2 / f_e$
0	24	10	14	196	19·60
1	13	20	−7	49	2·45
2	22	20	2	4	0·20
3	17	20	−3	9	0·45
4	19	20	−1	1	0·05
5	11	20	−9	81	4·05
6	18	20	−2	4	0·20
7	24	20	4	16	0·80
8	21	20	1	1	0·05
9	25	20	5	25	1·25
10	6	10	−4	16	1·60
	200	200			30·70

We call the total of the $(f_o - f_e)^2 / f_e$ terms χ^2. (χ is the Greek letter chi, pronounced "kye".)

So $\chi^2 = \sum \dfrac{(f_o - f_e)^2}{f_e}$ and in this case $\chi^2 = 30 \cdot 7$.

The smaller the discrepancies between expected and observed values, the smaller χ^2 will be. In particular if each discrepancy is zero, χ^2 will be zero. Thus the size of χ^2 is a measure of the goodness of fit of the theoretical model.

14.3. The χ^2-distribution

For any actual sample of data we would not expect the value of χ^2 to be exactly zero and more generally we would expect the actual value we obtained to depend on the chosen sample in much the same way as the sample mean and variance do. Like them the statistic χ^2 will have a distribution and we can talk about its sampling distribution just as we talked about the sampling distribution of the mean in Chapter 11. The distribution of the quantity χ^2 is a complicated one, because it will in general depend on the distribution of the model we are trying to fit, and to get any exact results is very difficult. However it turns out that, whatever the underlying theoretical distribution, with one small proviso that we shall meet later, the distribution of χ^2 is at least approximately of a standard form. Or rather, the distribution is one of a set of standard forms, for it depends on a parameter v, the number of degrees of freedom in the situation we are considering. Like the t-distribution, the χ^2-distribution has a different shape for different values of v. Also like the t-distribution, we will not give the exact form of the probability density function as it is extremely complicated. We have indicated the shape of the curve for a few values of v in Fig. 14.3 and the table on page 245 gives the probabilities associated with the distributions which we shall need.

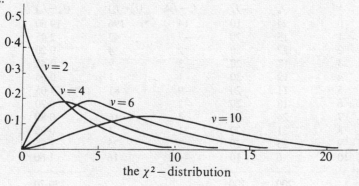

the χ^2 – distribution

Figure 14.3

Now let us use this information to analyse the above case where $\chi^2 = 30 \cdot 7$. We are really performing a hypothesis test, our null hypothesis being that the sample of waiting times does come from a population of waiting times which have a uniform distribution over the interval 0–10. The alternative hypothesis is that it comes from some other distribution. If the null hypothesis is true, χ^2 will have approximately a χ^2_v-distribution where v is the number of degrees of freedom. In this case the number of degrees of freedom is simply *the number of expected frequencies we can assign at will.* Here, since we have grouped our variable into 11 classes, it might seem that we could choose 11 expected frequencies arbitrarily; but we have the constraint that their sum must be 200, the total frequency, and so once we have decided on 10 of them, the eleventh is automatically determined. The number of degrees of freedom is $11 - 1 = 10$. In general for a variable divided into n classes the number of degrees of freedom will be $n - 1$, provided that the only constraint is that the total frequency is fixed. For every additional constraint one further degree of freedom is subtracted, as we shall see later and the number of degrees of freedom can equivalently be defined as the number of classes minus the number of constraints.

In this case our statistic has the χ^2_{10}-distribution. If we use a 5% significance level as usual, we wish to find the point in this distribution with 95% of the area to the left of it. As we can see from the table on page 245, 95% of the area of the χ^2_{10}-distribution lies to the left of 18·31.

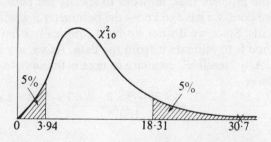

Figure 14.4

The value of 30·7 which we have obtained is well outside the 95% acceptance region, so we are forced to conclude that the model does not fit with the observed data very well, and we should alter our model. If we look at the way the value of 30·7 was built up, we see that a large proportion of it was caused by the very large discrepancy at $x = 0$, and this would suggest that we alter the model to lessen this.

Notice that in the test we have used the figure of 18·31 which corresponds to the 95% point of the distribution. If we suspected not that the distribution was not a good fit, but rather than someone had meddled with the results in order to make the fit appear too good, we could look at the 5% point, which is 3·94, instead, and say that if we obtained a χ^2-value which was less than 3·94 this would suggest that the results might have been "adjusted" as they fit rather too well.

14.4. Tests of goodness of fit for other distributions

The example we met above was a particularly simple one in that the theoretical model we were considering did not contain any unknown parameters. Let us consider a second example. On one particular Saturday last season the total number of goals scored in each of 150 football matches was counted with the following results.

x	0	1	2	3	4	5	6	7	8	9	10	11	12 or more
f	12	24	34	32	29	8	7	1	2	0	0	1	0

Can we find a distribution which will fit these results? As the distribution is discrete, we might try first the binomial or geometric or Poisson distributions, as these are the ones we have met most frequently (the distribution is clearly not uniform). In theory the number of goals scored in a match could be any whole number, and as the number of goals scored does not appear to decrease in a regular fashion as it would if the distribution were geometric, the Poisson distribution seems a reasonable first choice.

We have the problem that, in order to specify the particular Poisson distribution we need, we have to know the parameter μ which is the mean number of goals. Since we do not know theoretically what μ might be, our only method is to estimate it from the data. As we saw in question 2 of Exercise 11A, a "sensible" estimate to take is the sample mean $\bar{x}$.

Here we have

$$\bar{x} = \frac{1 \times 24 + 2 \times 34 + 3 \times 32 + 4 \times 29 + 5 \times 8 + 6 \times 7 + 7 \times 1 + 8 \times 2 + 11 \times 1}{150}$$

$$= 2{\cdot}8.$$

So we can assume that x has the $\mathscr{P}(2{\cdot}8)$ distribution and can proceed to calculate the expected frequencies of each value of x as $150P(x)$ where $P(x) = e^{-2{\cdot}8} \, 2{\cdot}8^x / x!$.

In the last section we said that the calculated value of χ^2 had approximately a χ^2- distribution subject to one proviso. This proviso is that the expected frequency in any class must not be too small. In fact if the expected frequency in any class is less than 5, the approximation cannot be assumed to be valid and in this case two or more classes must be combined in order to increase f_e.

In this example we have:

x	0	1	2	3	4	5	6	7	8	9	10	11 or more
$P(x)$	·061	·170	·238	·223	·156	·087	·041	·016	·006	·001	·001	·000
f_e	9·2	25·5	35·7	33·4	23·4	13·1	6·1	2·4	0·9	0·2	0·1	0·0

As we must have each value of $f_e \geqslant 5$, we must amalgamate the smaller classes into one larger one. If we form one class labelled "7 or more" we will still only have $f_e = 2\cdot4+0\cdot9+0\cdot2+0\cdot1+0\cdot0 = 3\cdot6$ for this class, so we are forced to form one class "6 or more".

Our final table for calculating χ^2 will look like this:

x	0	1	2	3	4	5	6 or more
f_o	12	24	34	32	29	8	11
f_e	9·2	25·5	35·7	33·4	23·4	13·1	9·7
$\dfrac{(f_o-f_e)^2}{f_e}$	0·85	0·09	0·08	0·06	1·34	1·99	0·17

$$\Sigma[(f_o-f_e)^2/f_e] = 4\cdot58$$

Hence $\chi^2 = 4\cdot58$. The number of degrees of freedom is the number of classes we have used less the number of constraints. Here we have used 7 classes. The constraints on f_e are that they must add to 150 and must have mean 2·8, so that we have 2 constraints and $v = 7-2 = 5$.

We find from looking at the tables for $v = 5$, that the 95% figure is 11·07, so the acceptance region is 0–11·07. Our value therefore falls in the acceptance region, and we accept the hypothesis that the Poisson distribution with mean 2·8 fits this sample of the total goals scored in football matches.

14.5. Fitting a normal curve

One distribution which we are very likely to want to fit is the normal distribution. Again this is a continuous curve, but any data to which we wish to fit it will necessarily be discrete. First then we must arrange the data into suitable classes so that we can compare the expected and observed frequencies within these classes. Of course we shall only be testing whether the distribution is a good fit to these classes and not how well it fits *within* the classes, so that we should not let the class widths be too wide. On the other hand we must ensure that the value of f_e is at least 5 in each class, so that we can use the χ^2-distribution.

We have to decide which of the family of normal distributions to fit to our data, which means that we have to estimate the two parameters μ and σ^2. We will use as estimates for these $\bar{x}$ and s^2 respectively.

Suppose we consider as an example the data we first met in Chapter 4 on the distances travelled by the snails, with the data grouped as there. If we look back to the appropriate chapters we will see that we had $n = 103$, $\bar{x} = 98\cdot5$ and $S = 52\cdot5$. (In this case we are justified in using S for s since the difference in dividing by 102 instead of 103 is negligible.) So if we want to fit a normal distribution we will naturally use $N(98\cdot5, (52\cdot5)^2)$.

Now we have to calculate the expected frequencies in each of our classes. The calculations are tedious, and this method is only used in practice when an exact result is needed. Often a quicker graphical method is used instead which, though approximate, is sufficiently accurate for most purposes. However, for the sake of completeness we give below the table for the calculation of χ^2 in our present example.

Class		z_1	z_2	$\Phi(z_1)$	$\Phi(z_2)$	$P(x_1 \leqslant x \leqslant x_2)$	f_e	f_o	
x_1	x_2	$\left(\dfrac{x_1 - 98\cdot5}{52\cdot5}\right)$	$\left(\dfrac{x_2 - 98\cdot5}{52\cdot5}\right)$			$\Phi(z_2) - \Phi(z_1)$			$\dfrac{(f_o - f_e)^2}{f_e}$
							$103P(x_1 \leqslant x \leqslant x_2)$		
*$-\infty$	37·5	$-\infty$	$-1\cdot16$	0·000	0·123	0·123	12·7	7	2·6
37·5	62·5	$-1\cdot16$	$-0\cdot69$	0·123	0·245	0·122	12·6	18	2·3
62·5	87·5	$-0\cdot69$	$-0\cdot21$	0·245	0·417	0·172	17·7	25	3·0
87·5	112·5	$-0\cdot21$	0·27	0·417	0·606	0·189	19·5	23	0·6
112·5	137·5	0·27	0·74	0·606	0·770	0·164	16·9	13	0·9
137·5	162·5	0·74	1·22	0·770	0·889	0·119	12·3	6	3·2
162·5	187·5	1·22	1·70	0·889	0·955	0·066	6·8	4	1·2
187·5	212·5	1·70	2·17	0·955	0·985	0·030	3·1	3	
212·5	237·5	2·17	2·65	0·985	0·996	0·011	1·1	0	1·3
237·5	262·5	2·65	3·12	0·996	0·999	0·003	0·3	2	
262·5	287·5	3·12	3·60	0·999	1·000	0·001	0·1	1	
287·5	∞*	3·60	∞	1·000	1·000	0·000	0·0	1	
							103	103	15·1

(braces group rows 187·5–287·5 for f_e totalling 4·6, and rows for f_o totalling 7, with $\dfrac{(f_o-f_e)^2}{f_e}$ = 1·3)

*Notice these two classes are extended so that the whole range of x is covered.

No. of classes used $= 8$.

No. of constraints $= 3$ $((a)$ total, (b) mean, (c) s.d. fixed$)$.

Hence $v = 8 - 3 = 5$.

As before the 95% value for χ^2_5 is $11\cdot07$.

Since $15\cdot1 > 11\cdot07$, we would reject the hypothesis of a normal distribution in this case.

EXERCISE 14A

1. Make up a table of between 600 and 1000 random numbers (it will help if several people each contribute shorter lists) or obtain such a table using a telephone directory, computer or the table at the back of the book. Test whether the distribution of (a) the single digits, (b) pairs of digits are what you would expect.

2. Tickets in a lottery had numbers in the range 1–10 000 inclusive. There were 250 winning tickets. The numbers on the winning tickets within successive ranges of 1000 numbers are given in the following table:

Range	Frequency of winning numbers	Range	Frequency of winning numbers
1–1000	16	5001–6000	22
1001–2000	24	6001–7000	26
2001–3000	32	7001–8000	30
3001–4000	38	8001–9000	18
4001–5000	18	9001–10000	26

Carry out a χ^2-test of goodness of fit to test, at the 5% level of significance, the hypothesis that all the tickets in the lottery had an equal chance of winning.

(M.E.I.)

3. Find how many times the figures 4 or 8 or both occur in the registration numbers of each of 100 cars and test whether the binomial distribution $B(3, 0.2)$ fits your results.

4. The following table shows the number of brown eggs occurring in 1000 boxes which each contain 6 eggs:

No. of brown eggs	0	1	2	3	4	5	6
Frequency	97	247	340	217	76	20	3

Estimate the probability that any particular egg is brown, and hence fit a suitable binomial distribution to the data, testing the goodness of fit.

5. On 40 consecutive weekdays the number of letters I received in my mail were as follows:

3, 5, 1, 1, 3, 3, 2, 4, 6, 1, 4, 0, 1, 3, 1, 2, 5, 4, 2, 1,
7, 4, 2, 5, 1, 3, 4, 1, 6, 5, 5, 1, 3, 4, 2, 5, 1, 3, 2, 4.

Calculate the mean and variance. Compare the frequencies with those which would be expected on the assumption of an appropriate Poisson probability model. Use a suitable test to investigate whether the difference could reasonably be attributable to chance. (S.M.P.)

6. The incidence of plumbing repairs in 78 council houses over a period of ten years is given in the table. Do the data support the view that there are good and bad tenants, or the view that there are lucky and unlucky tenants?

No. of repairs	0	1	2	3	4	5	6	7	8	9	10
No. of houses	3	13	16	16	10	9	3	5	1	1	1

(M.E.I.)

7.

No. of tests taken before passing	1	2	3	4	5	6	7	8	9
No. of drivers	75	53	38	12	10	3	7	2	0

Find the average number of tests $\bar{x}$ taken before a person passes the driving test and use $1/\bar{x}$ as an estimate of the probability p of a driver passing any particular test. Hence test the goodness of fit of a geometric distribution where
$$P(x) = (1-p)^{x-1}p, \qquad x = 1, 2 \ldots.$$

8. How well does the corresponding normal distribution fit the following data, showing the I.Q.s of 100 schoolchildren?

I.Q.	55–	65–	75–	85–	95–	105–	115–	125–	135–
No. of pupils	1	3	7	20	32	25	10	1	1

($\bar{x} = 100.45$, $s = 13.5$) (Each I.Q. is measured correct to one decimal place.)

9. The following data give the distribution of the numbers of species of a particular type of insect.

No. of species (x)	1	2	3	4	5	6	7	>7
Frequency (f)	88	31	14	8	4	3	2	0

Fit the following distribution to the data by using the ratio of Σfx to Σfx^2 in the sample in order to estimate q. The probability function to be fitted is $P(x) = K(1-q)^x/x$, $(x = 1, 2, 3, \ldots)$ with K a constant. Is the degree of agreement between observed and expected frequencies significant?

10. The following table shows the sizes of the gaps occurring between 100 successive vehicles on a particular road.

Gap (x) (to nearest 0.1 sec)	0–	5–	10–	15–	20–	25–	30–	35–40
Frequency	34	30	16	7	5	3	3	2

Show that $\bar{x} = 10$ and hence fit an exponential distribution $\mathscr{E}(\lambda)$, in which $f(x) = \lambda e^{-\lambda x}$ estimating λ by $1/\bar{x}$. How good is the fit?

14.6. Contingency tables

One of the great advantages of the χ^2-test is its wide range of application.

"You are all fair-weather footballers," said the manager of the Little Wickington 1st XI, "all right on a good pitch, but when it is rough going like today you have no idea!"

Jim, thinking that this was a bit unfair, looked up their results over the last two seasons, classifying the condition of the pitch for each match as "good", "medium" or "bad". His results are shown.

	wins	draws	losses
good	11	6	4
medium	12	7	7
bad	7	7	14

Here we have a list of observed frequencies for 9 groups. In order to make use of the χ^2-test we will evaluate the corresponding expected frequencies. Our null hypothesis must assume no connection between the result and the state of the pitch.

But is this to say that $\frac{1}{9}$ of the results should lie in each category? If we count up the totals we find that the team have been reasonably successful over the two seasons, and in fact the totals for wins, draws and losses are 30, 20, and 25 respectively. The frequencies of the different pitch conditions vary as well, the totals here being 21, 26 and 28 for good, medium and bad. In other words it would not be accurate to say that $\frac{1}{9}$ of the whole 75 matches would be expected to be draws on good pitches, as the team are less likely to draw and the pitch is less likely to be in a good condition. However if we do assume the null hypothesis, we would expect the ratios of wins: draws: losses to be 30:20:25, i.e. 6:4:5 or $\frac{6}{15}:\frac{4}{15}:\frac{5}{15}$, *whatever* the state of the pitch.

Hence the expected frequency of losses on a bad pitch is $\frac{5}{15}$ of matches on bad pitches $= \frac{5}{15} \times 28 = 9\cdot3$ (to 1 decimal place).

Similarly the expected frequency of a draw on a good pitch is $\frac{4}{15} \times 21 = 5\cdot6$.

In this way we can draw up a complete table of expected frequencies.

	f_o					f_e			
	W	D	L	Total		W	D	L	Total
G	11	6	4	21	G	8·4	5·6	7·0	21·0
M	12	7	7	26	M	10·4	6·9	8·7	26·0
B	7	7	14	28	B	11·2	7·5	9·3	28·0
Total	30	20	25	75	Total	30·0	20·0	25·0	75·0

Notice that the expected frequencies give the same row and column totals (and therefore grand total) as the observed frequencies. A simple rule, which we can verify, gives each expected frequency equal to

$$\frac{\text{(row total)} \times \text{(column total)}}{\text{grand total}}$$

We can now calculate the value of χ^2 in the usual way:

$$f_o - f_e$$

	W	D	L
G	2·6	0·4	−3·0
M	1·6	0·1	−1·7
B	−4·2	−0·5	4·7

$$\chi^2 = \frac{(2\cdot6)^2}{8\cdot4} + \frac{(0\cdot4)^2}{5\cdot6} + \frac{(3\cdot0)^2}{7\cdot0} + \frac{(1\cdot6)^2}{10\cdot4} + \frac{(0\cdot1)^2}{6\cdot9} + \frac{(1\cdot7)^2}{8\cdot7} + \frac{(4\cdot2)^2}{11\cdot2} + \frac{(0\cdot5)^2}{7\cdot5} + \frac{(4\cdot7)^2}{9\cdot3} = 6\cdot70$$

What about the number of degrees of freedom v? Suppose we think of assigning the expected frequencies to the table, and do it a row at a time. Once we have assigned the first two in row one, the third is fixed, as the row total is fixed. The same is true of row two. With row three, however, all the frequencies are fixed, as the column totals are fixed as well. So we can only assign $2 + 2 = 4$ expected frequencies arbitrarily and $v = 4$. Arguing similarly we can show that for a table with m rows and n columns, $v = (m-1)(n-1)$. (In general of course m need not equal n.) Here we have $\chi^2_4 = 6\cdot70$.

Looking up the 95 % value of χ^2_4 in tables we find that it is 9·49, so the result is not significant at 5 %. In other words Jim can go back and tell the manager that he has no real justification for his remarks. The result here is an interesting one, as the figures appear to give the opposite impression.

This type of table is known as a **contingency table.** Contingency tables are used to examine the possible association of two (or more) factors, as long as each factor gives rise to a certain number of categories like good, medium or bad; or win, draw and lose. They can be examined with the aid of the χ^2-test in the manner described above, provided of course each expected frequency is 5 or more. Should this not be the case, some of the categories must be combined until this is so.

The χ^2-test can be applied in the same way when one set of totals, e.g. the row totals, are predetermined. For a 2×2 table it is then equivalent to the two-tailed difference in proportions test (p. 186).

EXERCISE 14B

1. In a survey of smoking habits among 90 sixth-formers the following results were obtained:

	Boys	Girls
Non-smokers	16	24
Light smokers	20	10
Heavy smokers	14	6

Test whether there is a significant difference between the smoking habits of boys and girls.

2. Show that, for the 2×2 contingency table,

	B	not B
A	a	b
not A	c	d

$$\chi^2 = \frac{(ad - bc)^2 n}{(a+c)(a+b)(b+d)(c+d)}$$

where $n = a + b + c + d$.

3. Of 600 people who had been inoculated against the common cold, 40 caught severe colds within the next three months; of 400 people who had not been inoculated, 50 caught severe colds in the same period. Assuming that the samples were random, do these results provide any evidence of the effectiveness of inoculation? (M.E.I.)

4. 1890 car tyres were inspected in a survey of the maintenance habits of car owners. Examine the following data for evidence as to whether or not the spare wheel is exchanged regularly with the other wheels on cars:

	New	Part-worn	Worn	Badly worn
Wheels in use	350	902	288	73
Spare wheel	107	79	59	32

Write a short report on your analysis and conclusions. (M.E.I.)

5. Of 390 pairs of boots of type A issued to recruits, 68 had to be discarded after three months; of 210 pairs of boots of type B issued, 50 had to be discarded after three months. Use the χ^2-test to discuss the significance of the difference in the rates of wear of the two types of boot. Later, 200 pairs of type A and 180 pairs of type B were issued, and of these 36 of type A and 43 of type B had to be discarded after three months. Does the second trial, or the two trials taken together, show a significant difference in the rates of wear of the two types of boot? (M.E.I.)

6. Two-fifths of a sample of very overweight people were found to be neurotic, whereas only one-fifth of an equal number of people of normal weight were said to be so. Use the χ^2-test to find how large the sample size must be for the difference between the groups to be significant at the 5% level. Compare your answer with that of Exercise 13A, question 20.

7. In a 2×2 table it is advisable to use the 0·5 correction, by calculating χ^2 as

$$\sum \frac{(|f_o - f_e| - 0·5)^2}{f_e}.$$

The correction decreases the value of χ^2 and will only affect the conclusions from significant results where the expected values are small. It is known as *Yates' correction*. Calculate the uncorrected and corrected values of χ^2 from the table below and compare your results.

8	14
15	8

8. Examine the following data on the performance of candidates in the Sociology Honours Degree 1958–1964. Is there any evidence of significant differences in the performances of candidates from the two colleges?

College	I	IIA	IIB	III	Total
A	6	66	114	56	242
B	5	40	86	49	180

(M.E.I.)

9. A group of students were tested for introversion and neuroticism, to see whether the two characteristics were linked. The numbers of students in each category was as shown:

	extrovert	introvert
neurotic	2	5
stable	8	1

Why is the χ^2-test not applicable?

Can you work out the probability of 2 of the "neurotics" and 8 of the "stable" students being "extrovert", given that the total number of "extroverts" in the group is 10, and assuming that the two characteristics are independent?

How many more extreme arrangements than this one exist, where even fewer of the "neurotics" are "extrovert", given the same row and column totals? Work out the probabilities of such arrangements in the same way as above, and hence test the hypothesis that neuroticism and introversion are linked characteristics.

This technique is known as the *Fisher exact probability test*, and, like the χ^2-test, it is non-parametric.

Can you derive a formula for the probability of the general arrangement

$$\begin{array}{c|c} a & b \\ \hline c & d \end{array}$$ if the row and column totals are given?

REVISION EXAMPLES IV

1. It is desired to construct a significance test to choose between the following two hypotheses, as a result of one observation of a random variable X:

H_0: the variable X is uniformly distributed in the interval (1, 3).

H_1: the variable X is uniformly distributed in the interval (2·5, 3·5).

Construct a significance test such that the probability of accepting H_1 when H_0 is true is 0·05. For this test calculate the probability of accepting H_0 when H_1 is true. Construct a significance test with the property that the probability of accepting H_0 when H_1 is true is equal to the probability of accepting H_1 when H_0 is true. (Cam.)

2. A random variable x is uniformly distributed in the interval 0 to θ, where θ is an unknown positive quantity.

(i) Draw the graphs of the frequency function $f(y)$, and the distribution function (cumulative frequency function) $F(y)$ of the variable $y = x/\theta$.

(ii) Show that the statement $x/X_1 \leqslant \theta \leqslant x/X_2$ can be asserted with confidence $(X_1 - X_2)$.

(iii) If x is observed to be 39, at what significance level is the hypothesis $\theta = 40$ contradicted? (Cam.)

3. A cubical die with faces marked 1 to 6 is thrown n times. Show that on the hypothesis that the die is unbiased the chance that the face marked 4 will be uppermost not more than once is P, where

$$P = \left(\frac{n+5}{5}\right)\left(\frac{5}{6}\right)^n.$$

If $n = 40$ and the face marked 4 comes uppermost exactly once, test whether the hypothesis that the die is unbiased is contradicted:
(a) at the 1 % significance level,
(b) at the 0·1 % significance level. (Cam.)

4. The probability P_r that there will be r damaged tomatoes in a crate can be taken as $P_r = e^{-m}m^r/r!$, where m is the expectation of r. Over a large number of crates the value of m has been found to be 10. In the first crate from a new supplier the value of r was 4. Test whether this is significant evidence that the value of m for this supplier is less than 10; explain carefully the logic of your argument. (Cam.)

5. The diameter of a round peg is a normal variable with mean 1·48 cm and standard deviation 0·01 cm; the diameter of the hole in which the peg is supposed to fit is a normal variable with mean 1·50 cm and standard deviation 0·01 cm. Pegs and holes are paired at random. What proportion of pairs cannot be assembled because the radius of the peg is greater than that of the hole with which it is paired? If the hole is more than 0·04 cm larger than the peg, then the peg is so loose that it falls out. What proportion of assembled pairs are too loose to hold together? (Cam.)

6. If a grocer claims "All eggs sold in this shop are fresh", the finding of one bad egg is enough completely to disprove his claim. Why can no certain proof or disproof be obtained if he claims "90% of eggs sold in this shop are fresh"? I believe that in fact only 50% of the eggs are fresh but I will not dispute the claim unless less than 70% of a sample are fresh. Determine n so that for any sample of n or more eggs there is more than 0·9 chance of disputing the claim when my belief (50% fresh) is actually correct. (Cam.)

7. A certain manufacturer claims that at least 80% of motorists fit his products to their cars. In a random sample of 100 cars 75 are found to be so fitted. Test whether this evidence justifies the rejection of his claim at the 5% significance level. Test similarly the evidence given by finding 300 fitted cars in a random sample of 400. Find also the smallest value of n(to the nearest 10) for which the finding of only $\frac{3}{4}n$ fitted cars in a random sample of n cars justifies the rejection of the manufacturer's claim at the 5% significance level.

(M.M.E.)

8. The following table gives the distribution of systolic blood pressure, in millimetres of mercury, for 254 male workers aged 30 to 39 years.

Blood pressure (mid-interval value)	85	95	105	115	125	135	145	155	165	175
Number of men	1	2	12	47	77	65	32	13	4	1

Calculate the mean and the standard deviation of these pressures. Determine the standard error of the mean, and calculate 99% confidence limits for the mean blood pressure of the population of which the above group is a random sample. (M.E.I.)

9. Two resistors are both made up by joining ten coils in series, so that the resistance of each resistor is the sum of the resistances of the ten coils. The coils are chosen at random from a population with standard deviation 0·1 ohm. Obtain limits within which the difference between the resistances of the two resistors will lie in 95 cases out of 100. (Cam.)

10. Let $\bar{x}$ be the mean of a random sample of n observations from a normal population having an unknown mean and a known variance σ^2. It is proposed to test whether the population mean has some assumed value μ_0 by rejecting this assumption if $\dfrac{\sqrt{n}}{\sigma}|\bar{x} - \mu_0| \geqslant C_\alpha$, where C_α is a constant depending on the significance level, α per cent, of the test. Find the value of C_α corresponding to $\alpha = 15$ and $\alpha = 3$. Let $P(\mu)$ be the probability that the assumed value μ_0 is rejected when the population mean has a general value μ. Show that

$$P(\mu) = 1 - \Phi\left[\frac{(\mu_0 - \mu)\sqrt{n}}{\sigma} + C_\alpha\right] + \Phi\left[\frac{(\mu_0 - \mu)\sqrt{n}}{\sigma} - C_\alpha\right],$$

where $\Phi(x)$ is the cumulative distribution function of the standardized normal distribution. Sketch the graphs of $P(\mu)$ for $\alpha = 15$ and $\alpha = 3$. (J.M.B.)

11. A random variable X is uniformly distributed on the interval $(0, \beta)$. An independent random sample of n observations is drawn from this distribution. The mean of the sample is $\bar{x}$ and the largest observation is $x_{(n)}$. Write down an expression for the probability that all the observations are less than x, and deduce that the probability density function of $x_{(n)}$ is nx^{n-1}/β^n for $0 < x < \beta$. Show that $2\bar{x}$ is an unbiased estimator of β. Show also that $((n+1)/n)x_{(n)}$ is another unbiased estimator of β. Calculate the sampling variances of these two estimators, and use these variances to compare the accuracies of the two estimators. (J.M.B.)

12. A large number of seeds of a white-flowering plant are mixed by machine with twice the number of seeds (of about the same size and weight) of a red-flowering plant. If there is full germination in beds each sown with 100 seeds of the mixture, what are the mean and standard deviation (s.d.) of the theoretical population of white-flowering plants per bed, assuming the machine produces a randomized mixture? A count of the white-flowering plants in ten such beds (with full germination) gave the figures: 29, 30, 30, 26, 38, 25, 34, 30, 31, 32. How many of these lie within (a) one s.d., (b) two s.d. of the mean of the theoretical population? Can you conclude from this anything about the validity of the hypothesis of randomization? Test this hypothesis in another way by finding the mean of the given figures. What is your opinion about the hypothesis on the evidence of the data? (S.M.P.)

13. (a) Explain briefly what is meant by the statement that "a sample mean differs significantly from some assumed value".

    (b) Discuss the merits of using the sample median, sample mode, or sample mean to estimate a population mean, and also the role of sampling distributions in deciding which statistic should be used. (J.M.B.)

14. The percentages of damaged cells in eight suspensions were determined by a differential staining technique and by phase-contrast microscopy with the following results:

Suspension	1	2	3	4	5	6	7	8
Staining	8·3	13·8	16·4	11·7	9·4	31·5	20·7	12·3
Phase-contrast	7·8	12·6	14·3	10·8	9·9	22·4	15·8	13·5

Plot the values obtained with the two techniques as ordinate and abscissa respectively and use the paired sample t-test to compare the mean percentages. How far do you consider that this comparison is (a) valid, (b) helpful?   (Cam.)

15. A university maintenance officer buys 10 light bulbs from a firm which claims that its bulbs have an average life of 1000 hr. He measures the life x hours of the bulbs in the sample and finds $\Sigma x = 9602$, $\Sigma x^2 = 9\,240\,000$. Does this provide evidence against the firm's claim? He then buys a sample of 8 bulbs from a second firm and tests them. This sample gives $\Sigma x = 7923$, $\Sigma x^2 = 7\,892\,500$. Use the two-sample t-test to determine whether there is a significant difference between the bulbs of the two firms. What distributional assumptions are made in your analysis?   (Cam.)

16. A random sample from a normal distribution with unknown mean and variance $\sigma^2 = 4$, yields a symmetric two-sided 95 % confidence interval $(0·73, 1·71)$ for the mean. Calculate the sample mean, and the number of observations in the sample. A further random sample, from a second normal distribution with variance $\sigma^2 = 4$, has mean 1·78. Discuss whether there is significant evidence at the 5 % level that the means of the two distributions are different:
 (i) from the above facts alone,
 (ii) if it is also known that the two samples are of the same size.
Show that whatever the size of the second sample the information in the data is insufficient to demonstrate a difference between the means which is significant at the 1 % level.   (J.M.B.)

17. It was suspected that candidates from schools with small sixth forms did not have a fair chance of getting into a certain university. A study was therefore made of the 60 applicants for one faculty; the following list gives the size of sixth form for each applicant's school and whether they were successful (s) or unsuccessful (u) in gaining a place:

204s	157s	245s	89s	301u	161s	60s	196s	196u	113s
48s	238s	302u	335s	150s	67u	301s	206s	45u	78u
215s	335u	128u	56u	126s	223s	88s	215u	302s	28u
60s	124u	255u	137s	192u	229s	68s	136s	343s	45s
39u	109s	38u	218u	207s	310s	229s	218u	178s	199s
70u	77u	192u	302s	146u	69s	39u	94u	124s	84s

Taking a "small" sixth form to be one with fewer than 100 pupils, analyse the data statistically and discuss whether they give support to the suspicion of unfairness. Present the information in the table in a suitable graphical form to show the degree of success in schools with sixth forms of various sizes.
    (S.M.P.)

18. In a particular experiment when reading a certain scale, the last digit has to be estimated. The table below gives the distribution of 400 readings by an observer. The 400 readings have been chosen at random from a large number of readings by that observer and have been classified according to the magnitude of the estimated digit. The table shows that the digits 0 and 8 appear more frequently than the others. Determine whether or not the observer is showing a systematic bias towards these digits.

Digit	0	1	2	3	4	5	6	7	8	9	
Observed frequency	60	31	28	44	37	54	26	28	62	30	(M.E.I.)

Relationships between Two Variables

15.1. Introduction

In earlier chapters we have been mainly concerned with experiments and results concerning one variable at a time; in this one we wish to consider two variables and the relationships between them. We could ask if the cigarette sales in any year have any connection with the number of deaths from lung cancer in that year, or if there is some connection between how well someone plays rugby and how good he is at French.

For an example suppose we consider the First Division of the Football League in the 1969–70 season. In the final table the teams are arranged in order according to the number of points they scored in the season, where each game won scores 2 points and each game drawn scores 1 point. The table also records the number of games played, won, lost and drawn, and the number of goals scored by the team and against them. We might be interested in considering the relationship between the number of points gained by a team and the number of goals scored by them. Is it true, as the table seems to suggest, that the more goals a team scores the more points they win, or do some teams score points largely as a result of their defensive play? Equally is it true that the more points a team scores the fewer goals they have scored against them? Often anyone interested in doing football pools will also want to know whether there is any obvious relationship between the position of a team in the table and the number of occasions on which the team draws during the season.

Take in particular these last two variables: the points scored by a team and the number of drawn games they play. We know how to consider either variable separately and find its mean and variance, but now we are interested in how they interact. Are they both large or small together or, conversely, is one large when the other is small? Do they appear to satisfy any simple relationship, say that of a straight-line law?

213

FINAL LEAGUE TABLE

SEASON 1969–70

Division I

	P.	W.	L.	D.	Goals For	Ag.	Pts.
Everton	42	29	5	8	72	34	66
Leeds United	42	21	6	15	84	49	57
Chelsea	42	21	8	13	70	50	55
Derby County	42	22	11	9	64	37	53
Liverpool	42	20	11	11	65	42	51
Coventry City	42	19	12	11	58	48	49
Newcastle United ...	42	17	12	13	57	35	47
Manchester United ...	42	14	11	17	66	61	45
Stoke City	42	15	12	15	56	52	45
Manchester City ...	42	16	15	11	55	48	43
Tottenham Hotspur ...	42	17	16	9	54	55	43
Arsenal	42	12	12	18	51	49	42
Wolverhampton W. ...	42	12	14	16	55	57	40
Burnley	42	12	15	15	56	61	39
Nottingham Forest ...	42	10	14	18	50	71	38
West Brom. Albion ...	42	14	19	9	58	66	37
West Ham United ...	42	12	18	12	51	60	36
Ipswich Town	42	10	21	11	40	63	31
Southampton	42	6	19	17	46	67	29
Crystal Palace	42	6	21	15	34	68	27
Sunderland	42	6	22	14	30	68	26
Sheffield Wednesday ...	42	8	25	9	40	71	25

Or does there appear to be little if any connection between them? We have two objectives in asking questions such as these: first to examine any relationship that may exist and see if we can suggest any reason for it; and second, given such a relationship, to use values of one variable in order to predict values of the other. We will concern ourselves with the first objective in this chapter and leave prediction until the next.

We should first like to draw a diagram to illustrate the combined behaviour of any two variables. The most usual way of doing this is to draw a **scatter diagram.** The value of one variable is taken as the x-coordinate and that of the other as the y-coordinate with respect to the usual pair of axes and a point marked to represent each pair of corresponding values of the two variables.

Trouble arises when two or more pairs of values are identical; the second and subsequent occurrences are usually shown as circles drawn round the point representing the values.

If we look at the three diagrams we can see roughly what we expect. The dots on the diagram showing "points" and "goals for" appear to lie fairly close to a line of positive slope; the larger the number of points, the larger the number of goals scored. Just the reverse is true for the

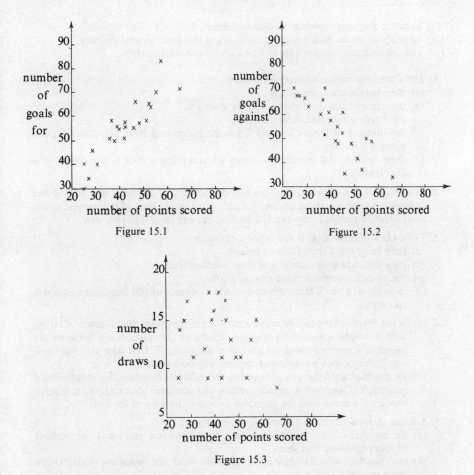

Figure 15.1

Figure 15.2

Figure 15.3

diagram of "points" and "goals against" where the dots lie fairly close to a line of negative slope. The diagram for "points" and "draws" has no definite form; high number of points seems to be little related to either high or low number of draws.

The next set of examples gives further pairs of variables which you can examine and for which you can draw scatter diagrams.

INVESTIGATIONS III

Examine the relationship between some of the following pairs of variables, or any other variables in which you are interested. You should write a report on each investigation giving:

(i) details of the data collection,

(ii) a scatter diagram showing the two variables,
(iii) an analysis of the data using the methods of this and the next chapter,
(iv) any conclusions you are able to draw about the relationship.

1. For a random sample of people consider:
   (a) their height and their weight,
   (b) their height and the cube root of their weight,
   (c) their arm length and their shoe size,
   (d) the number of letters in their Christian name and their surname (maiden name if a woman),
   (e) their vertical and horizontal errors when aiming a dart at the centre of a dart board.

2. Toss two dice a number of times (20 or more) and consider any two of the following variables: score on the first die, score on the second die, largest score, total score, difference between the scores. Repeat for a different pair.

3. For a class of boys or girls at a school, examine:
   (a) their height and their father's height,
   (b) their distance from school and their journey time,
   (c) their marks in two or more examinations,
   (d) their time to run 100 metres and to add a column of 100 five-figure numbers correctly.

4. (a) In the Physics laboratory make a pendulum by suspending a piece of string with a weight attached to one end. Examine the relationship between the length of the string l and the period of the swing T. Consider also log l and log T. Any other suitable experiment will do instead.
   (b) In the Biology laboratory examine the relation between the length l and breadth b of leaves of a certain species, and compare your results for various species. Examine also the variables $l \times b$ and the area of the leaf.

5. Use the *Annual Abstract of Statistics* to examine:
   (a) the number of car accidents in each of the last few years and the number of cars registered that year,
   (b) the number of attendances at cinemas and the number of television licences sold,
   (c) the number of school pupils and the number of school teachers, or any other suitable variables.

6. When shopping examine:
   (a) the prices of a random selection of goods at each of two supermarkets,
   (b) the price of a commodity which is sold in various weights or sizes (such as paint or flower-pots), and the weight or size. Consider also the price per unit weight and the weight.

7. Use an R.A.C. or A.A. book to make a random selection of towns; examine their distance from London as the crow flies and their minimum distance by road. Or, using a railway guide, compare rail travelling times with direct distances.

In each of the above categories you will probably be able to think of other suitable examples.

15.2. The coefficient of correlation

It is easy to say of our scatter diagrams that the plotted points "appear to lie fairly close to a line", or "seem to be little related". What we would now like to find is a numerical measure of the relationship. When we were considering a single variable, we measured its variability by a quantity which we called its *variance* and which, given a sample of n values of the variable, we estimated by

$$s^2 = \frac{\Sigma(x-\bar{x})^2}{n-1}.$$

Now with two variables x and y we can estimate the variance of each and denote these by s_x^2 and s_y^2 respectively. These quantities measure the variability of each variable "within itself"; it might seem sensible if we measured the variability "between the variables" by a quantity which we call the **covariance** and estimate by

$$s_{xy} = \frac{\Sigma(x-\bar{x})(y-\bar{y})}{n-1}.$$

(Note that this is simply an extension of our variance formula, for the covariance of x with itself is

$$s_{xx} = \frac{\Sigma(x-\bar{x})(x-\bar{x})}{n-1} = s_x^2)$$

We have used the degrees of freedom, $n-1$, in our expression for s_{xy}, since we are usually concerned with estimating a population covariance from a set of sample data. Examination of s_{xy} shows that, if x and y are large together and small together, $(x-\bar{x})$ and $(y-\bar{y})$ will be either both positive or both negative and s_{xy} will consist of a sum of positive terms; conversely if x is large when y is small and vice versa, s_{xy} will consist of a sum of negative terms. If there is no such definite relationship between x and y, s_{xy} will consist of some positive and some negative terms which will tend to cancel. It appears that s_{xy} will give some measure of the relationship between x and y.

There is a difficulty that s_{xy} will depend on the units in which x and y are measured. Suppose we take one of the examples from Investigations III (opposite) and consider the covariance of the heights of boys and their fathers. First we measure the heights of the fathers in metres and calculate s_{xy}; then we change all their heights to centimetres and compute the new covariance. Since we have multiplied all the y-values, and therefore $\bar{y}$ as well, by 100, we will also have multiplied s_{xy} by 100. In order to reflect truly the variability between x and y, we would like a value which is unaffected by the units in which we measure either of them. In the example above, as well as multiplying s_{xy} by a factor of 100, it is clear that we will multiply s_x^2 by a factor of 10 000 and therefore s_x by a factor of 100. We can see that if we considered not s_{xy} but $\frac{s_{xy}}{s_x s_y}$, we would have a measure which was unaffected by the units of both x and y.

We call this new measure the **correlation coefficient** and usually write it as r. So we have

$$r = \frac{s_{xy}}{s_x s_y} = \frac{\Sigma(x-\bar{x})(y-\bar{y})}{\sqrt{\{\Sigma(x-\bar{x})^2 \Sigma(y-\bar{y})^2\}}}$$

since the factors $(n-1)$ cancel.

15.3. The calculation of r

In Chapter 6 we saw that the easiest way of evaluating $\Sigma(x-\bar{x})^2$ was to expand it and use the form $\Sigma x^2 - n\bar{x}^2$. We can find a corresponding expression for $\Sigma(x-\bar{x})(y-\bar{y})$:

$$\Sigma(x-\bar{x})(y-\bar{y}) = \Sigma(xy - \bar{x}y - \bar{y}x + \bar{x}\bar{y}) = \Sigma xy - \bar{x}\Sigma y - \bar{y}\Sigma x + n\bar{x}\bar{y}$$
$$= \Sigma xy - n\bar{x}\bar{y} - n\bar{x}\bar{y} + n\bar{x}\bar{y} = \Sigma xy - n\bar{x}\bar{y}.$$

We can use these results to simplify our calculation of r. For example let us calculate r for two variables taking the following 5 pairs of values;

x	1	2	3	4	5
y	4	7	10	13	16

We have $\Sigma x = 15$ $\therefore \bar{x} = 3$, $\Sigma y = 50$ $\therefore \bar{y} = 10$

$\Sigma x^2 = 55 \Rightarrow \Sigma(x-\bar{x})^2 = 55 - 5.3.3 = 10$

$\Sigma y^2 = 590 \Rightarrow \Sigma(y-\bar{y})^2 = 590 - 5.10.10 = 90$

$\Sigma xy = 180 \Rightarrow \Sigma(x-\bar{x})(y-\bar{y}) = 180 - 5.3.10 = 30$ $\therefore r = 30/\sqrt{900} = +1$

This example illustrates a general result which is fundamental for the correlation coefficient. If we had plotted a scatter diagram for these values, we would have seen that all the points lie exactly on a line of positive slope (and it is not hard to see from the values that the line has equation $y = 3x + 1$). It is true in general, and not hard to prove, that r always takes the value $+1$ for sets of points lying exactly on such a line. Correspondingly r always takes the value -1 for sets of points which lie on a line whose slope is negative. All other values of r will be between these two extremes and so we have the result $-1 \leqslant r \leqslant 1$. (You will find this proof as an exercise later.) Thus our value r really measures how close sets of points are to lying on a straight line. Variables for which r is near zero have little or no *linear* relationship; they may well have some definite relationship, as we shall see later, but this will be of some form other than a straight-line law. Our three examples with the football table illustrate three types of relationship, and we should be able to guess at least roughly the values of r in each of these cases. Variables for which $r > 0$ are said to be *positively correlated* (and this implies increasing values of x are associated with increasing values of y); those for which $r < 0$ are said to be *negatively correlated*; and those for which $r = 0$ *uncorrelated*. If $r = \pm 1$ the correlation is said to be *perfect*.

We can investigate the correlation between some of the pairs of variables in Investigations III (p. 216). If either (or both) of the variables we are working with is continuous or takes many discrete values, we may find it necessary to group our data into class intervals in the usual way. We can estimate r from the table of grouped data just as we did for the variance in Chapter 6 by replacing the formula for r by

$$\frac{\Sigma f(x - \bar{x})(y - \bar{y})}{\sqrt{\{\Sigma f(x - \bar{x})^2 \Sigma f(y - \bar{y})^2\}}}$$

where each x and y is the mid-point of one of the classes and the corresponding f is the number of values which fall in that class. For example, we could group the football data giving "points" and "goals for" into classes as in the table below.

As we have seen earlier, r is unaffected by the units in which the x- and y-values are measured; it is also clear that, since all calculations involve deviations from the mean, r will not be affected by the origin about which the x- and y-values are measured. So there is no reason why we should not choose a convenient origin and units for x and y for the purposes of calculation. The full results are set out below.

			GOALS FOR										
		Class	30–36	37–43	44–50	51–57	58–64	65–71	72–78	79–85			
		Mid-pt.	33	40	47	54	61	68	75	82			
		Coded	-3	-2	-1	0	1	2	3	4			
class	mid-pt.	coded									f	fx	fx^2
25–29	27	-4	2	1	1						4	-16	64
30–34	32	-3		1							1	-3	9
35–39	37	-2			1	2	1				4	-8	16
40–44	42	-1				4					4	-4	4
45–49	47	0				2	1	1			4	0	0
50–54	52	1					1	1			2	2	2
55–59	57	2						1		1	2	4	8
60–64	62	3									0	0	0
65–69	67	4							1		1	4	16
		f	2	2	2	8	3	3	1	1	22	-21	119
		fy	-6	-4	-2	0	3	6	3	4	4		
		fy^2	18	8	2	0	3	12	9	16	68		

POINTS

$\Sigma fx = -21$ $\bar{x} = -21/22$ $\Sigma fx^2 = 119$ $\Sigma f(x - \bar{x})^2 = 119 - 21^2/22 \simeq 98 \cdot 95$
$\Sigma fy = 4$ $\bar{y} = 4/22$ $\Sigma fy^2 = 68$ $\Sigma f(y - \bar{y})^2 = 68 - 4^2/22 \simeq 67 \cdot 27$
$\Sigma fxy = 2(-3)(-4) + 1(-2)(-4) + 1(-1)(-4) + 1(-2)(-3) + \ldots = 69$ $\Sigma f(x - \bar{x})(y - \bar{y}) = 69 + 84/22 \simeq 72 \cdot 82$
$r = \dfrac{72 \cdot 82}{\sqrt{\{98 \cdot 95 \times 67 \cdot 27\}}} \simeq 0 \cdot 89$

15.4. Drawing conclusions from the sample correlation coefficient

When for example we find the correlation coefficient r between the heights of people and their weight, what we are really doing is calculating the coefficient for a sample (hopefully random) of values from a very large population of values. We should like to be able to use the value of r we have obtained in the sample to discuss the correlation between the twŏ variables in the whole population. Even though we obtain a large positive value of r it could be that we have just been looking at an odd sample and that the correlation in the population, which we shall call ρ (the Greek letter rho), is still 0 (although of course if our sample size n was fairly large and our sample a truly random one, we should be surprised if the obtained value of r was too far from that of ρ).

What we need is a significance test. Various methods exist for testing whether we are likely to have obtained a given r from a population with given ρ. We will only consider the case when the null hypothesis is that $\rho = 0$. On this hypothesis it can be assumed that the function

$$\frac{r}{\sqrt{1-r^2}}\sqrt{n-2}$$

has a t-distribution with $n-2$ degrees of freedom, provided the values are approximately normally distributed. Hence with our null hypothesis $H_0 : \rho = 0$ and our alternative $H_1 : \rho > 0$, we would reject H_0 for large positive values of

$$t = \frac{r}{\sqrt{1-r^2}}\sqrt{n-2}.$$

For example, if we obtained $r = 0.28$ for a sample of 25 values, then we would have

$$t = \frac{0.28}{\sqrt{[1-(0.28)^2]}}\sqrt{23} = 1.40.$$

From the t-tables the critical value at the 5% level for 23 degrees of freedom is 1·71 (approximately) and so, since $1.40 < 1.71$, we would accept the null hypothesis of no correlation in the population. You can probably see how we could have carried out the corresponding two-tailed test.

Suppose we have established a significant correlation between two variables. It would then be tempting to seek an explanation for the relationship we have discovered. This is thorny ground though, and statisticians are often content to say that the variables are linearly related and not go further and say that an increase in one *causes* an increase (or decrease) in the other. It is a fact that two variables may increase or decrease together, and hence be positively correlated, even though no casual interaction is at work. A classic example occurred in America some years ago when a survey showed a large positive correlation between the number of crimes committed in a given year and the number of church attendances in the year. The fact was that while both variables happened to be increasing over the years in question, probably neither "caused" the other to increase. In other cases, for example, the heights of fathers and sons, a causal effect may be more likely.

EXERCISE 15A

1. Calculate r for the following six pairs of values of x and y: (1, 2), (2, -1), (1, 2), (0, 5), (3, -4), ($1\frac{1}{2}$, $\frac{1}{2}$). How would you explain the value you obtain?

2. Calculate r for the four pairs of values (4, 5), (4, -5), (-4, 5), (-4, -5) lying on the circle $x^2 + y^2 = 41$. Find another set of 4 pairs of values lying on the circle for which $r > 0.95$. Comment on your results.

3. The following table shows the Intelligence Quotient (I.Q.) and the mark obtained in an examination for each of 10 children. Evaluate the product moment correlation coefficient between the I.Q.s and the examination marks.

I.Q.:	145	100	100	112	138	133	123	116	127	106
Mark:	75	35	52	36	80	90	71	54	52	55

(J.M.B.)

(Note the *product moment correlation coefficent* is another name for r which we have simply called the *correlation coefficient*.)

4. Calculate the coefficient of correlation between the weight of the heart and the weight of the liver in mice, using the following data. What do you conclude from this analysis?

Weight of organs (in units of 0·01 g)

Heart	20	16	20	21	26	24	18	18
Liver	230	126	203	241	159	230	140	242

(M.E.I.)

5. In a study of the quality of imported bacon, the following observations were obtained of the leaness of the carcass (x) and the palatability (y):

x	1·8	2·6	3·4	5·2	5·9	6·8	8·3	11·5	12·0	13·0	13·1	13·8
y	5·7	5·0	7·8	8·9	10·4	10·2	10·2	9·7	7·5	4·8	7·0	5·5

Plot the observations and calculate the correlation coefficient between x and y using working means of 7 for x and 8 for y. Comment on the value of the correlation coefficient. (Cam.)

6. Show that, for a set of pairs of values of x and y which satisfy the relationship $y = mx + c$, $r = +1$ if m is positive and -1 if m is negative.

7. Guess the value of r for the two remaining pairs of variables used in §15.1. Find how accurate your guess was by calculating r in each case. (You may group the data if you wish.)

8. The frequency table given below shows the results of giving a test in French and a test in English to a class of 30 boys. Each test is marked out of 10. Calculate the coefficient of correlation and comment on your result.

Marks in English	2	3	4	5	6	7	8	9	10
9							1	1	
8					1			1	
7						1		1	
6		1		2		2			
5			2	1	1			2	
4					2		2	1	
3	1	2			1	2			
2				1			1		

Marks in French

(Cam.)

9. The bivariate frequency table given below shows data collected in order to determine whether or not there is any association between the times a particular species of animal takes in searching for food and in eating it. The searching time X and the eating time Y are given in seconds and the mid-points of the grouping intervals are shown in the table. Calculate the product moment correlation coefficient and give verbal conclusions from the analysis. The values of X and Y should be coded with origin (20, 57·5).

X	Y 37·5	57·5	77·5	97·5	117·5	137·5	Total
20	11	30	24	6	—	1	72
35	4	16	9	4	1	2	36
50	1	2	—	1	—	2	6
65	—	—	—	2	—	—	2
80	—	1	—	—	—	—	1
95	—	—	1	—	—	—	1
Total	16	49	34	13	1	5	118

(M.E.I.)

10. Show that $s^2_{x+y} = s^2_x + 2r\,s_x s_y + s^2_y$ and $s^2_{x-y} = s^2_x - 2r\,s_x s_y + s^2_y$.

11. Show that

$$\Sigma\left(\frac{x-\bar{x}}{s_x} + \frac{y-\bar{y}}{s_y}\right)^2 = 2(n-1)(1+r)$$

and find a similar expression for

$$\Sigma\left(\frac{x-\bar{x}}{s_x} - \frac{y-\bar{y}}{s_y}\right)^2.$$

Deduce that $-1 \leqslant r \leqslant 1$.

12. If the correlation coefficient of x and y is r, and the standard deviations of x and y are s_x and s_y respectively, show that the correlation between $x+y$ and x is given by

$$\frac{(rs_y + s_x)}{\sqrt{\{s^2_x + 2rs_x s_y + s^2_x\}}}$$

Hence, using the results of question 7, find the correlation between points and games won for the football teams.

13. By expanding $\Sigma[(x-\bar{x}) + \lambda(y-\bar{y})]^2$ as a quadratic function of λ, deduce that $s^2_{xy} \leqslant s^2_x s^2_y$ and hence that $-1 \leqslant r \leqslant 1$.

14. For a sample of 14 values r was calculated to be 0·50. Would you reject (at the 5% level) the hypothesis that ρ for the whole population was zero against (i) $\rho > 0$, (ii) $\rho \neq 0$?

15. If the data in question 4 were considered to be a typical sample from all mice, would you reject the hypothesis that, for the whole population, the correlation between weight of heart and liver was zero against the alternative that it was greater than zero?

16. Use the data of question 9 to test whether there is evidence of a non-zero correlation between the searching and eating times for the particular species considered.

17. Show that if $r/\sqrt{1-r^2} = x$ then $r = \sin \theta$ where $\tan \theta = x$. Hence, for $n = 5$, 10, 17, 22, 32, 62, find the least positive value of r for which you would reject $\rho = 0$ against $\rho \neq 0$ for a sample of that size. (Use the 5% level of significance.)

15.5. Ranking methods

Often in comparing two variables the actual values that they take are not in fact very crucial; what is more important is the order in which the values are placed.

Two judges, Mr. Fair and Mr. Wise, are judging a class at a dahlia show. They award the following marks to the exhibits A–F.

	A	B	C	D	E	F
Mr. Fair	19	41	85	27	63	72
Mr Wise	46	45	54	48	51	49

Is there evidence to show that the judges agree in their marking? It is clear that what matters here is not the actual marks awarded (for each marking system is probably peculiar to the judge using it) but rather the order in which each places the exhibits. So we code the marks for each judge giving 1 to the exhibit to which each gave the highest mark, 2 to the second highest and so on. This is the process of *ranking* which was introduced in the section on non-parametric tests in 13.6.

	A	B	C	D	E	F	
Mr. Fair	6	4	1	5	3	2	(x)
Mr. Wise	5	6	1	4	2	3	(y)

If the judges had been in perfect agreement the results would be

1	2	3	4	5	6
1	2	3	4	5	6

arranged in some order. If they had been in total disagreement

1	2	3	4	5	6
6	5	4	3	2	1

We would like a measure indicating how near to these two extremes our two judges are, and giving a value of $+1$ for perfect agreement and -1 for total disagreement. We have already met one measure which will do this, for perfect agreement implies that the ranks are related by the straight-line equation $y = x$ and total disagreement by the straight line $y = 7 - x$, so we could use the correlation coefficient r. When applied to ranks r is known as *Spearman's coefficient* (of rank correlation). We will write it as r_s.

In our example we had:

$\Sigma x = \Sigma y = 21 \qquad \therefore \bar{x} = \bar{y} = \frac{7}{2}$

$\Sigma x^2 = \Sigma y^2 = 91 \quad \therefore \Sigma(x-\bar{x})^2 = \Sigma(y-\bar{y})^2 = 91 - 6 . \frac{7}{2} . \frac{7}{2} = \frac{35}{2}$

$\Sigma xy = 87 \qquad \therefore \Sigma(x-\bar{x})(y-\bar{y}) = 87 - 6 . \frac{7}{2} . \frac{7}{2} = \frac{27}{2}$

$\therefore r_s = \frac{27}{2} / \frac{35}{2} = \frac{27}{35} \simeq 0 \cdot 77.$

If, in general, n objects are ranked and there are no tied ranks,

$\Sigma x = \Sigma y = \frac{1}{2}n(n+1) \qquad \bar{x} = \bar{y} = \frac{1}{2}(n+1),$

$\Sigma x^2 = \Sigma y^2 = \frac{1}{6}n(n+1)(2n+1) \quad \Sigma(x-\bar{x})^2 = \Sigma(y-\bar{y})^2 = \frac{1}{12}n(n^2-1).$

Hence
$$r_s = \frac{\Sigma xy - \frac{1}{4}n(n+1)^2}{\frac{1}{12}n(n^2-1)}$$

When there are tied ranks this new formula should not be used. In this case we can either use a modified form which is derived in Exercise 15B, or calculate r_s from the original formula.

An alternative measure which is sometimes used instead of r_s is *Kendall's coefficient*, which we shall call r_k though it is often called τ. In this measure every possible pair of exhibits is considered; each pair is allotted a score of $+1$ if they are in the same order in both x- and y-rankings and a score of -1 if they are in the reverse order. The total of the scores is then obtained. The easiest way to calculate the coefficient is to arrange one of the variables so that the ranks are in the correct order as below:

	C	F	E	B	D	A	
Mr. Fair	1	2	3	4	5	6	(x)
Mr. Wise	1	3	2	6	4	5	(y)

Then a total is formed by starting at the left-hand y-number and counting $+1$ for each number greater than itself, and -1 for each number less than itself among the other y-ranks. This gives the score for the first y-rank. Similarly the score for the second y-rank is obtained by repeating the process but using only those y-ranks to the right of it in the table. We can verify that the total so formed is the same as the total in our original definition.

For our example

y	1		3		2		6		4		5	
score	5	+	2	+	3	+	(-2)	+	1	+	0	Total 9

In order to convert this total to a coefficient taking value $+1$ for perfect agreement, we must divide it by the maximum possible total. It is not hard to see that if there are no tied ranks this must be just the number of pairs, 15 in our example and $\frac{1}{2}n(n-1)$ in the general case, since if the agreement were perfect each pair would contribute a score of $+1$.

So in our example we have $r_k = \frac{9}{15} = 0.6$. Notice that this is not equal to r_s. The coefficients give different numerical values but both indicate a reasonable measure of agreement between the judges.

In calculating r_k in cases where there are tied ranks, any pair which are tied in one or both of the rankings should contribute a score of 0 to the total score. The denominator should also strictly be changed from $\frac{1}{2}n(n-1)$ to $\sqrt{\{[\frac{1}{2}n(n-1) - U][\frac{1}{2}n(n-1) - V]\}}$ where U is the number of tied pairs in the x-ranking and V the number of tied pairs in the y-ranking; but this latter correction is only important when the proportion of tied pairs is large, as otherwise the expression is approximately equal to $\frac{1}{2}n(n-1)$.

Of course we should like to be able to carry out a significance test before drawing any conclusions from ranked sample data. For large values of n we can test r_s approximately by treating it as an ordinary correlation coefficient and using the method of the last section. For small n, and in general for r_k, the tests are more complicated and by and large we shall simply regard rank correlation coefficients as giving quickly-calculated approximate measures of the relationship between two variables.

The significance tests for both r_s and r_k are, like those in §§13.5–13.7, non-parametric tests.

EXERCISE 15B

1. Here are the marks of eight candidates in physics and mathematics. Rank the results and hence find a rank correlation coefficient between the two sets of marks.

Candidate	A	B	C	D	E	F	G	H
Physics mark	43	51	29	82	73	40	38	58
Maths mark	62	65	50	71	32	67	42	49

(S.M.P.)

2. In a beauty competition two judges place the contestants in the following orders:
   (i) CEDFAGIJBH;
   (ii) FGDAICHEJB.
   Find a coefficient of rank correlation between the two orders. (S.M.P.)

3. Show that an alternative form for **Spearman's** coefficient of rank correlation between two sets of n ranks (when none of the ranks are tied), is

$$1 - \frac{6\Sigma d^2}{n(n^2 - 1)},$$

where d is the difference between the ranks of corresponding values.

4. Three judges in a beauty competition place the six finalists in the following orders:

Competitor	A	B	C	D	E	F
First Judge	3	6	1	5	2	4
Second Judge	2	5	3	6	4	1
Third Judge	1	4	2	6	5	3

Find a coefficient of rank correlation between the first two judges and also between the first and third judges. Do the findings of the first judge agree better with those of the second or third judge? (Cam.)

5. The death rates R_1 and R_2 (deaths per year per thousand of the population) in a certain district from pneumoconiosis and cancer of the lung, grouped according to age, are as follows:

Age-group	20–25	25–30	30–35	35–40	40–45	45–50	50–55	55–60	60–65	65–
R_1	3·7	3·2	6·1	8·4	12·1	11·3	16·4	18·9	18·3	19·8
R_2	2·3	2·4	4·6	4·6	5·2	6·4	6·2	7·8	8·8	8·6

Calculate a rank correlation coefficient between the two sets of death rates and comment on your result. (Cam.)

6. Calculate a rank correlation coefficient for one (or more) of the pairs of variables considered in the football data and compare your result with the corresponding unranked coefficient already calculated.

7. An educational psychologist obtained scores by 9 university entrants in 3 tests (A, B and C). The scores in tests A and B were as follows:

Entrant	1	2	3	4	5	6	7	8	9
A score	8	3	9	10	4	9	6	4	5
B score	7	8	5	9	10	6	3	4	7

Calculate a coefficient of rank correlation between the two sets of scores. The coefficient obtained between the A and C scores was 0·71 and that between the B and C scores was 0·62. What advice would you give the psychologist if he wished to use less than three tests? (Cam.)

8. A set of n values are ranked. If one set of m values are given the same rank, show that the variance of the set of ranks is $(n(n^2-1)-m(m^2-1))/12$. Deduce that r_s can be written as

$$(\Sigma xy - \tfrac{1}{4}n(n+1)^2)/\sqrt{[(\tfrac{1}{12}n(n^2-1)-\Sigma'\tfrac{1}{12}m(m^2-1))(\tfrac{1}{12}n(n^2-1)-\Sigma''\tfrac{1}{12}m(m^2-1))]}$$

where the summation Σ' is taken over all sets of tied ranks in the x-variable and Σ'' over all those in the y-variable. Verify the result from the data of question 7 (first calculating r_s in the ordinary way if you have not already done so).

Fitting Straight Lines to Data

16.1. Introduction

In the last chapter we discovered how to investigate the extent to which two variables satisfy a linear relationship by considering their correlation coefficient. Suppose we find that the corresponding values do lie reasonably well on a straight line; how would we set about finding its equation? We could of course draw it in by eye, if an approximate result is all we need. But is there a more exact method? If we could find the line that best fits the pairs of points, we could find an approximate formula for one variable in terms of the other, and we could then use it to estimate values of the one for different values of the other. This, remember, was the second of our objectives in the last chapter.

Notice that there is an immediate difference here from the previous chapter; there the two variables were considered equally, and no distinction was made between them; here we talk about finding a formula for "one variable in terms of the other", and immediately introduce the idea of one dependent and one independent variable. This is an important point, and it is always as well to consider at the beginning of any problem which variable is to be considered dependent and which independent. For example:

Geoffrey is doing an experiment in the Physics laboratory to examine the relationship between the length of a spring and the weight of an object attached to it. He wants to find a formula giving approximately the length of the spring in terms of the weight. He attaches 6 known weights in turn to the end of the spring and measures the corresponding lengths. When he plots his 6 results on a scatter diagram he finds that they lie approximately on a straight line.

In this kind of situation it is usually clear which should be the dependent variable y and which the independent variable x. Here y will be the length of the spring and x the weight attached. In fact, provided we assume the weights to be sufficiently accurate, x is not really a random variable, as we know the exact values it takes. Geoffrey is only concerned with finding a formula expressing y in terms of x, which is called a (linear) **regression** of y on x.

A rather different example is provided in the following situation:

The 15 pupils on a certain computer course were each given an aptitude test before the course started and a final test at the end of it. Their marks in the two tests were:

Aptitude test	59	63	81	40	55	29	43	72	49	98	80	67	35	51	70
Final test	40	62	71	24	45	22	39	58	31	86	75	76	42	51	59

We might want to ask two different types of question about these data.

(i) If we had a 16th pupil who scored 60 in the aptitude test but missed the final test, what is the best estimate of his mark?

(ii) If the pass mark in the final test is 50, what is the likely mark in the aptitude test of someone who just passes the final test? Let us first plot the marks on a scatter diagram. We will call the aptitude test mark x and the final test mark y, though the choice is purely arbitrary.

Calculation shows that for these values $r \simeq 0.90$. The variables are approximately linearly related and we could best answer each question by fitting a straight line to the data.

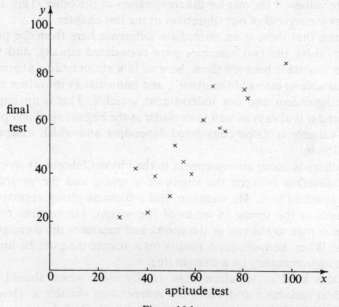

Figure 16.1

The two questions are asking rather different things. Question (i) is asking for a formula giving the final mark in terms of the aptitude mark, which is treating y as the dependent and x as the independent variable. Fitting the best straight line in this situation is a regression of y on x. Question (ii) on the other hand is asking for a prediction of the aptitude mark in terms of the final mark and treats x as dependent and y as independent. This is a regression of x on y.

There is then a difference between our two examples. In the former, only y is a true random variable, and we are only interested in a regression of y on x. In the second example both x and y are random variables and we can sensibly talk about regressions of y on x and x on y; which of these we use in any situation will depend on the question we wish to answer. As we shall see, the assumptions we make in the two situations will be different and so in general will be the straight lines we obtain.

16.2. The regression of y on x

Let us consider the regression of y on x. We assume that y is approximately a linear function of x and that the relationship would be exact if it were not for an "error" term. Here the "error" term may represent a true measurement error as in Geoffrey's example or may just be the

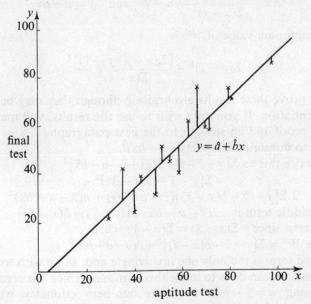

Figure 16.2—Values of error term

difference between an actual value of y and the average of the population of y-values for that x, as in our second example. Though the assumptions are rather different in the two examples it turns out mathematically that the two are equivalent. We assume further that if we could obtain the whole population of y-measurements which could correspond to any particular x, the average value of the "error" in all of them would be zero, that the errors are independent and have equal variance.

This means that we can write y as $y = a + bx + e$, where e is a small error term whose average value is zero. Note that $e = y - a - bx$ and therefore e measures the amount by which y is above or below the assumed line. The intercept a and the gradient b of the line are unknown, and we want to make the best estimates we can of them. We shall call these estimates $\hat{a}$ and $\hat{b}$. Best estimates will clearly make e as small as possible, or, since we have many pairs of points, will make some function of all the es as small as possible. We could choose just to minimize the sum of the es, but this would allow the possibility of a large positive and a large negative e cancelling, which would clearly not give us a good line. Instead, therefore, we decide to minimize Σe^2.

This involves finding the minimum of $W^2 = \Sigma(y - a - bx)^2$. As we show, the minimizing values $\hat{a}$ and $\hat{b}$ are given by the two equations:

$$\hat{b} = \frac{\Sigma(x - \bar{x})(y - \bar{y})}{\Sigma(x - \bar{x})^2} = \frac{s_{xy}}{s_x^2} \quad \text{and} \quad \bar{y} = \hat{a} + \hat{b}\bar{x}$$

and the minimum value of W^2 is

$$\Sigma(y - \bar{y})^2 - \frac{[\Sigma(x - \bar{x})(y - \bar{y})]^2}{\Sigma(x - \bar{x})^2}$$

We shall prove these results algebraically though they may be obtained by differentiation. If you only wish to use the results, you may want to omit the proof and go straight to the next paragraph.

We wish to minimize $W^2 = \Sigma(y - a - bx)^2$.

We can write this as $\Sigma[y - \bar{y} - b(x - \bar{x}) + \bar{y} - a - b\bar{x}]^2$

$$= \Sigma[y - \bar{y} - b(x - \bar{x})]^2 +$$
$$2\,\Sigma[y - \bar{y} - b(x - \bar{x})](\bar{y} - a - b\bar{x}) + n(\bar{y} - a - b\bar{x})^2$$

but the middle term is $2(\bar{y} - a - b\bar{x})\,\Sigma[(y - \bar{y}) - b(x - \bar{x})]$

which is zero since $\Sigma(x - \bar{x}) = \Sigma(y - \bar{y}) = 0$.

Therefore $W^2 = \Sigma[y - \bar{y} - b(x - \bar{x})]^2 + n(\bar{y} - a - b\bar{x})^2$.

The second term is the only one involving a and, as we wish to minimize W^2, the best we can do is to make this term zero (since it is certainly $\geqslant 0$) by choosing $a = \bar{y} - b\bar{x}$. Therefore our best estimates will satisfy $\hat{a} = \bar{y} - \hat{b}\bar{x}$ which is the second of the equations above. Now we have to

chose $\hat{b}$ to minimize $W^2 = \Sigma[(y-\bar{y})-b(x-\bar{x})]^2$. We can write this as
$$W^2 = b^2\Sigma(x-\bar{x})^2 - 2b\Sigma(x-\bar{x})(y-\bar{y}) + \Sigma(y-\bar{y})^2$$
$$= \Sigma(x-\bar{x})^2\left(b - \frac{\Sigma(x-\bar{x})(y-\bar{y})}{\Sigma(x-\bar{x})^2}\right)^2 + \Sigma(y-\bar{y})^2 - \frac{[\Sigma(x-\bar{x})(y-\bar{y})]^2}{\Sigma(x-\bar{x})^2}$$

by 'completing the square'. Again the best we can do to minimize W^2 is to make the first term zero, that is to let $\hat{b} = \dfrac{\Sigma(x-\bar{x})(y-\bar{y})}{\Sigma(x-\bar{x})^2}$.

Then we see that this makes the minimum value of W^2 equal to $\Sigma(y-\bar{y})^2 - \dfrac{[\Sigma(x-\bar{x})(y-\bar{y})]^2}{\Sigma(x-\bar{x})^2}$. (Those familiar with calculus will recognize that we could have obtained these results by finding $\dfrac{\partial W}{\partial a}$ and $\dfrac{\partial W}{\partial b}$ and equating them to zero.)

Notice the similarity between the formula for $\hat{b}$ and the one we met for the correlation coefficient r in the last chapter. We will find the relationship between them in a later section. Notice also that, since the second equation is $\bar{y} = \hat{a} + \hat{b}\bar{x}$, the point $(\bar{x}, \bar{y})$ lies on our best line, which means that the line passes through the 'centre of gravity' of the set of points.

These equations enable us to answer the first of our questions about the computer tests, for we can now find the regression line of y on x for the marks. We have:

$\Sigma x = 892$ $\bar{x} \simeq 59\cdot5$ $\Sigma(x-\bar{x})^2 \simeq 5085\cdot7$
$\Sigma y = 781$ $\bar{y} \simeq 52\cdot1$ $\Sigma(x-\bar{x})(y-\bar{y}) \simeq 4739\cdot5$
$\hat{b} = 4739\cdot5/5085\cdot7 \simeq 0\cdot93,$ $\hat{a} \simeq 52\cdot1 - 59\cdot5 \times 0\cdot93 \simeq -3\cdot35.$

We can thus say that the best straight line for this problem is

$$y = 0\cdot93x - 3\cdot35$$

We can now mark in the fitted line on our scatter diagram. The easiest way to draw in the line is first to mark the point $(\bar{x}, \bar{y})$ which must lie on the line and then any other convenient point on it, say where $x = 100$, and then join up the two as on the next page.

We can now estimate the final mark for an aptitude mark of 60 either by reading it off from the graph or evaluating it from the equation of the line. We find that the mark is approximately 52·5. What exactly does this answer represent? It is not an exact value for y as $\hat{a}$ and $\hat{b}$ are only estimates of a and b. Also the error term for this particular case is unknown and cannot be taken into account. Although it is unlikely that any actual y will have a zero error term, we assumed at the outset

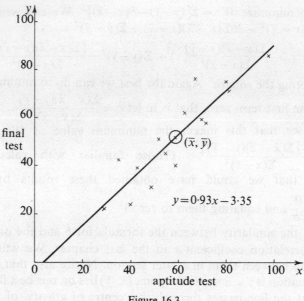

Figure 16.3

that the average error term for all values of y for any given x is zero. So we can say that the value we obtain is an estimate of the average value of y for that particular x.

16.3. The regression of x on y

In answering our second question about the tests, we meet just the opposite situation and wish to find the best line expressing x in terms of y. We can write this as $x = c + dy + f$ where f is another "error" term which we shall also assume has average value zero. We can write $f = x - c - dy$ and so f measures the amount by which the point lies to the left or the right of the fitted line. The intercept c and the gradient d are unknowns and we wish to find the best estimates we can of them, $\hat{c}$ and $\hat{d}$.

The obvious thing to do this time is to minimize $T^2 = \Sigma(x - c - dy)^2$, and we can clearly obtain the solutions by interchanging x and y in the equations we had before. This gives us

$$\hat{d} = \frac{\Sigma(x - \bar{x})(y - \bar{y})}{\Sigma(y - \bar{y})^2} = \frac{s_{xy}}{s_y^2} \quad \text{and} \quad \bar{x} = \hat{c} + \hat{d}\bar{y}$$

and the minimum value of T^2 is $\Sigma(x - \bar{x})^2 - \dfrac{[\Sigma(x - \bar{x})(y - \bar{y})]^2}{\Sigma(y - \bar{y})^2}$

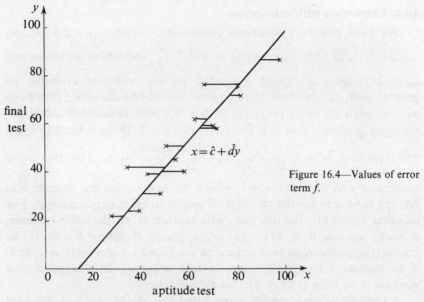

Figure 16.4—Values of error term f.

We can now easily show that the required line in our case is
$$x = 14 \cdot 2 + 0 \cdot 87 y$$
and hence the estimated aptitude mark corresponding to a final mark of 50 is about 58.

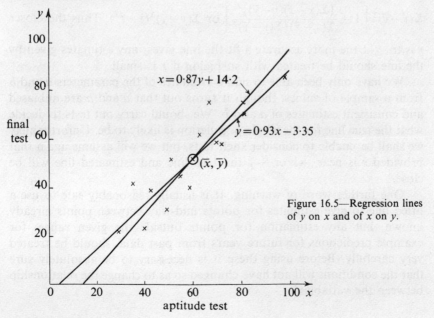

Figure 16.5—Regression lines of y on x and of x on y.

16.4. Connection with correlation

We have now two equations connecting x and y; $y = \hat{a} + \hat{b}x$ and $x = \hat{c} + \hat{d}y$ $\left(\text{which we can rewrite as } y = \dfrac{x}{\hat{d}} - \dfrac{\hat{c}}{\hat{d}}\right)$ and which we use in any given circumstance depends on what we are trying to estimate. In general, as in our example, the two lines will not be the same; they both pass through the same point $(\bar{x}, \bar{y})$ but will only coincide if they have the same gradient, that is if $\hat{b} = 1/\hat{d}$, i.e. $\hat{b}.\hat{d} = 1$. If we substitute for $\hat{b}$ and $\hat{d}$ we have $\hat{b}\hat{d} = \dfrac{\Sigma(x - \bar{x})(y - \bar{y})}{\Sigma(x - \bar{x})^2} \cdot \dfrac{\Sigma(x - \bar{x})(y - \bar{y})}{\Sigma(y - \bar{y})^2} = r^2$. The two lines coincide if and only if $r = \pm 1$, which we saw in the last chapter was just the condition for the two sets of points to lie exactly on a line. For all other values of r the two lines will be distinct. At the other extreme, if $r = 0$, we will have $\Sigma(x - \bar{x})(y - \bar{y}) = 0$ and therefore $\hat{b} = \hat{d} = 0$. So the two regression equations reduce to $y = \bar{y}$ and $x = \bar{x}$ respectively, and, if we assume a straight-line law in this situation, knowledge of one variable is no help to us in predicting the other.

A measure of the accuracy of our fitted line, say of y on x, is provided by the minimum sum of squares of the errors. We saw earlier that this was equal to $\Sigma(y - \bar{y})^2 - \dfrac{[\Sigma(x - \bar{x})(y - \bar{y})]^2}{\Sigma(x - \bar{x})^2}$ and we can rewrite it as $\Sigma(y - \bar{y})^2 \left[1 - \dfrac{(\Sigma(x - \bar{x})(y - \bar{y}))^2}{\Sigma(x - \bar{x})^2 \Sigma(y - \bar{y})^2}\right]$ or $\Sigma(y - \bar{y})^2(1 - r^2)$. Thus the closer r is to ± 1 the more accurate a fit the line gives; any estimates given by the line should be treated with suspicion if r is small.

We have only been able to make estimates of the parameters a and b from a sample of values, though it turns out that $\hat{a}$ and $\hat{b}$ are unbiased and consistent estimates of a and b. We should carry out tests to decide what the true line for the whole population is likely to be. Unfortunately we shall be unable to consider such tests, but we will assume again that provided r is near $+1$ or -1 the true line and estimated line will be close.

One further word of warning. It is usually reasonably safe to use a fitted line to give estimates for points mid-way between points already known, but any estimation for points outside the given range, for example predictions for future years from past data, should be treated very carefully. Before using these it is necessary to be absolutely sure that the conditions will not have changed so as to change the relationship between the variables.

16.5. Fitting other curves

Using the technique by which we derived the equations of the lines in the earlier sections, it is possible to fit a variety of other curves to data. The technique is known as the **method of least squares** as it involves minimizing a sum of squares of errors.

Many of the curves we would like to fit, however, can be derived neatly from the straight-line formula. Suppose we had a set of pairs of points x and y which appeared to lie on a curve of the form $y = ab^x$ and we wished to estimate the parameters a and b. We can rewrite the curve as $\log y = \log a + x \log b$ and, provided we make suitable assumptions about the error, we are back in a familiar situation. We can fit a straight line to the set of points x and $\log y$, estimate $\alpha = \log a$ and $\beta = \log b$, and then derive our estimates of a and b as e^{α} and e^{β}. You can probably find a similar method if the curve predicted is of the form $y = ax^b$. You will find some examples of fitting other curves in Exercise 16.

EXERCISE 16

1. American tourists visiting Great Britain.

Year	Year with 1953 as origin	No. of tourists (thousands)	Amount they spent (£ millions)
	t	x	y
1953	0	185	36·1
1954	1	203	38·8
1955	2	240	42·5
1956	3	255	46·3
1957	4	266	47·7
1958	5	325	57·0
1959	6	360	64·0

Use the above data to plot graphs of (i) x against t and (ii) y against t, and in each case draw by eye the straight line which best fits the points plotted. Obtain the equations of your straight lines in the form $x = m_1 t + c_1$, $y = m_2 t + c_2$. Extend the table to show the estimated number of tourists and the amount which they will spend in each of the years 1960 and 1961. Discuss the reliability of these estimates.

(J.M.B.)

2. The length x mm of a metal rod was measured at various temperatures $t°C$, giving the following results:

t:	105·0	110·0	115·0	120·0	125·0	130·0
x:	200·1	200·8	201·8	202·0	204·3	205·1

By means of a linear regression equation estimate the length of the rod when the temperature is $122·5°C$.

(J.M.B.)

3. The number of students graduating in physics in each of the years 1960 to 1964 is shown in the following table:

Year	1960	1961	1962	1963	1964
Number	885	921	963	950	1143

Calculate the line of regression of the number of new physicists on the year and show the results of your calculation graphically. Estimate the number of new physicists graduating in 1968. Discuss the reliability of your estimate.

(M.E.I.)

4. George has started playing golf and has played one round nearly every Saturday morning for the last ten weeks. His scores are given in the following table:

1	2	3	4	5	6	7	8	9	10
96	93	—	89	86	83	—	79	82	76

Estimate his score for the weeks he did not play and his score in week 15. Why would you suspect the last estimate?

5. The unadjusted index x of industrial production in Great Britain for the twelve months of 1965 is given in the following table (1958 index 100):

Jan.	132	Feb.	137	Mar.	137	April	129	May	136	June	131
July	120	Aug.	116	Sept.	133	Oct.	138	Nov.	140	Dec.	133

Taking a month as a unit of time and $t = 0$ for January 1965, calculate:
(a) the means and variances of x and t,
(b) the covariance of x and t,
(c) the equation of the line of regression of x on t,
(d) the correlation coefficient of x and t.
Comment briefly on these results. (M.E.I.)

6. Show that for two pairs of observations x_1, y_1 and x_2, y_2 the regression lines of both y on x and x on y are simply the straight line passing through both points.

7. If the regression line of y on x is $y = a + bx$, find the regression lines of:
(i) z on x where $z_i = py_i + q$;
(ii) y on w where $w_i = rx_i + s$.

8. The following table gives, in coded form, the breadths and lengths of 1306 human heads.

Frequencies of heads
X (length)

	−4	−3	−2	−1	0	1	2	3	4	5	Total
−4			2	2	1	1	2		1		9
−3	1		5	5	15	8	4	1	1		40
−2	2		3	21	34	48	41	18	3	1	171
−1			4	21	57	92	110	57	19	5	1	366
0	1		2	17	53	93	116	62	27	12	1	384
1			2	3	12	39	65	57	40	12	2	232
2				1	4	9	17	24	12	9	1	77
3					1	1	5	10	6		2	25
4							2				2
Total	4	18	70	177	291	360	231	109	38	8	1306

The length X is coded with 18·85 cm as origin and the breadth Y is coded with 15·15 cm as origin. The interval of grouping is 0·4 cm in both cases. Calculate the average length and the average breadth of the heads, and the equation of the line of regression of breadth on length. (M.E.I.)

9. If it is known that the relation between x and y should be of the form $y = mx$ except for an error term, estimate m by the method of least squares from a sample of values of x and y.

10. Fit a curve of the form $y = ax^2 + b$ to the following values and show the fitted curve graphically.

x	3	4	5	6	7
y	10	18	27	36	50

11. The sales of a glass company over the period 1957–66 are given below in units of £50,000. Show graphically that the company is in a period of rapid expansion, and use the method of least squares to fit a curve of the form $y = cd^x$, where c and d are constants.

Year	1957	1958	1959	1960	1961	1962	1963	1964	1965	1966
Sales	132·5	142·6	143·8	171·0	203·8	203·8	228·5	217·8	294·2	303·5

Plot the fitted curve to the sales data. (M.E.I.)

12. Three variables, x, y and z are observed together on seven occasions and the following values obtained:

Occasion		1	2	3	4	5	6	7	Total	Mean
Variable	x	4	8	11	9	14	16	12	74	10·57
	y	3	9	8	7	14	12	11	64	9·14
	z	17	12	4	2	5	1	−4	37	5·29

Show that the corrected sum of squares (c.s.s.) of x

$$\left[\text{i.e. } \sum_{i=1}^{7}(x_i - \bar{x})^2 \right]$$

and the corrected sum of products (c.s.p.) of y with z

$$\left[\text{i.e. } \sum_{i=1}^{7}(y_i - \bar{y})(z_i - \bar{z}) \right]$$

are approximately 95·71 and −95·29.

Given that the corrected sum of squares of y and z are 78·86 and 299·43, and that the corrected sum of products of x with y and x with z are 78·43 and −127·14, obtain the regression equations of x on y, x on z and y on z. Find the values of x and y predicted by these equations for:

(a) $z = 5$, (b) $z = 15$, and also the values of x predicted when
(c) $y = 9·23$, (d) $y = 6·05$.

Comment on the results of these calculations. (Cam.)

REVISION EXAMPLES V

1. In the following diagrams the points represent observed pairs of values of x and y; the lines shown are drawn as attempts to summarize the relationships and not as the results of known theoretical arguments. Comment on the aptness of the captions and drawn lines.

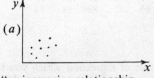

"an increasing relationship between y and x"

"no significant regression of y on x"

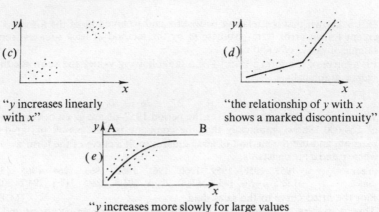

(c)

"y increases linearly
with x"

(d)

"the relationship of y with x
shows a marked discontinuity"

(e)

"y increases more slowly for large values
of x and tends to the asymptote AB"   (Cam.)

2. A random sample of twelve aluminium die-castings was chosen and measurements made of their tensile strength and hardness. The results (to two
significant figures, in appropriate units) were as follows:

| Tensile strength: | 29 | 35 | 37 | 30 | 34 | 34 | 32 | 30 | 32 | 35 | 37 | 32 |
| Hardness: | 53 | 70 | 84 | 55 | 79 | 71 | 77 | 64 | 69 | 58 | 79 | 62 |

Draw a scatter diagram for these data. Calculate the product moment correlation coefficient. Discuss what is suggested by the scatter diagram and correlation coefficient about the relationship between tensile strength and hardness for
these aluminium die-castings.   (J.M.B.)
(You should carry out an appropriate significance test.)

3. There are ten finalists in a beauty competition. Two judges, P and Q, separately
place the finalists in order of merit as shown in the following table.

Finalist	A	B	C	D	E	F	G	H	I	J
Order by P	4	1	9	5	6	10	2	7	8	3
Order by Q	7	3	10	2	4	8	1	6	9	5

Calculate a coefficient of rank correlation and state what you can deduce
from your result.   (Cam.)

4. The sales in units of £1000 of a certain firm for the years 1960 to 1966 are given
in the table. Use these data to estimate the sales in 1970.

Year	1960	1961	1962	1963	1964	1965	1966
Sales	65·5	70·0	73·0	79·0	82·0	82·5	88·5

Comment on the validity of the estimate.   (M.E.I.)

5. Price index numbers of home-grown wheat (x) and cereals (y) at ten successive
quarters are given in the following table:

| x | 84 | 88 | 102 | 101 | 84 | 75 | 80 | 83 | 100 | 99 |
| y | 78 | 81 | 95 | 94 | 92 | 85 | 82 | 88 | 98 | 83 |

Find:
(a) the covariance of x and y,
(b) the line of regression of y on x.
Explain briefly the distinction between the line of regression which you have
calculated and that for x on y. Under what circumstances would they be
sensibly the same?   (M.E.I.)

6. Eight boys' marks in mathematics and physics in an examination were as follows:

Mathematics:	20	35	40	67	81	52	26	55
Physics:	50	60	52	62	68	57	56	65

Calculate the mean and standard deviation of the marks in each subject. Explain which you consider to be a better mark, 52 in mathematics or 57 in physics. Exhibit the results on graph paper with, for each boy, the mathematics mark as the x- and the physics mark as the y-coordinate. Comment on the pattern obtained. Another boy in the same group scored 53 in physics but missed the other examination. Obtain as fairly as possible the mark he would have expected to score in mathematics. (S.M.P.)

7. The proportion of technologists in the combined total numbers of scientists and technologists and the rates of economic growth per capita per cent per annum are given for twelve countries. Investigate the association between the rate of growth and the technological effort. (The data may be ranked.)

Country	Proportion of technologists	Rate of growth	Country	Proportion of technologists	Rate of growth
Ireland	0·57	1·6	Denmark	0·73	2·7
U.K.	0·58	2·1	Belgium	0·77	2·1
Italy	0·62	5·0	Austria	0·78	5·2
Canada	0·65	0·4	Netherlands	0·80	3·7
France	0·68	3·1	Sweden	0·81	2·9
U.S.A.	0·71	1·1	Norway	0·83	2·1

(M.E.I.)

8. The following table gives the value observed for the dependent variable y which is subject to experimental error, when an independent variable is adjusted to have the value x.

x	0	1	2	3	4
y	0·51	0·99	1·61	2·02	2·48

Plot the values of y against x and ascertain which one of them appears to be grossly in error. Using the remaining four values only, find the equation of the regression line of y on x. What value does the regression equation predict for the aberrant observation? Assuming that each y-value is the mean of 36 independent determinations each with the same standard deviation 0·22 and that the predicted value is effectively the true mean value of the variable y, find the probability of such a mean of observations being as aberrant as or more aberrant than the one in question. (Cam.)

9. Use the method of least squares to fit a parabolic curve of regression of y on x of the form $y = a + bx + cx^2$ to the following pairs of values:

x	-2	-1	0	1	2
y	10	7	$2\frac{1}{2}$	$1\frac{1}{2}$	3

(M.E.I.)

10. Two judges are asked to rank three wines for quality, neither judge being permitted to rank two wines as equal. If one of the judges ranks the wines in a random order, find the probability that the absolute value of the rank correlation coefficient exceeds 0·6. (Either Spearman's or Kendall's coefficient may be used.) (Cam.)

Tables

Table 1

THE NORMAL DISTRIBUTION FUNCTION

z	$\Phi(z)$	z	$\Phi(z)$	z	$\Phi(z)$	z	$\Phi(z)$
0·00	$0·5000_{40}$	0·40	$0·6554_{37}$	0·80	$0·7881_{29}$	1·20	$0·8849_{20}$
·01	$·5040_{40}$	·41	$·6591_{37}$	·81	$·7910_{29}$	·21	$·8869_{19}$
·02	$·5080_{40}$	·42	$·6628_{36}$	·82	$·7939_{28}$	·22	$·8888_{19}$
·03	$·5120_{40}$	·43	$·6664_{36}$	·83	$·7967_{28}$	·23	$·8907_{18}$
·04	$·5160_{39}$	·44	$·6700_{36}$	·84	$·7995_{28}$	·24	$·8925_{19}$
0·05	$0·5199_{40}$	0·45	$0·6736_{36}$	0·85	$0·8023_{28}$	1·25	$0·8944_{18}$
·06	$·5239_{40}$	·46	$·6772_{36}$	·86	$·8051_{27}$	·26	$·8962_{18}$
·07	$·5279_{40}$	·47	$·6808_{36}$	·87	$·8078_{28}$	·27	$·8980_{17}$
·08	$·5319_{40}$	·48	$·6844_{35}$	·88	$·8106_{27}$	·28	$·8997_{18}$
·09	$·5359_{39}$	·49	$·6879_{36}$	·89	$·8133_{26}$	·29	$·9015_{17}$
0·10	$0·5398_{40}$	0·50	$0·6915_{35}$	0·90	$0·8159_{27}$	1·30	$0·9032_{17}$
·11	$·5438_{40}$	·51	$·6950_{35}$	·91	$·8186_{26}$	·31	$·9049_{17}$
·12	$·5478_{39}$	·52	$·6985_{34}$	·92	$·8212_{26}$	·32	$·9066_{16}$
·13	$·5517_{40}$	·53	$·7019_{35}$	·93	$·8238_{26}$	·33	$·9082_{17}$
·14	$·5557_{39}$	·54	$·7054_{34}$	·94	$·8264_{25}$	·34	$·9099_{16}$
0·15	$0·5596_{40}$	0·55	$0·7088_{35}$	0·95	$0·8289_{26}$	1·35	$0·9115_{16}$
·16	$·5636_{39}$	·56	$·7123_{34}$	·96	$·8315_{25}$	·36	$·9131_{16}$
·17	$·5675_{39}$	·57	$·7157_{33}$	·97	$·8340_{25}$	·37	$·9147_{15}$
·18	$·5714_{39}$	·58	$·7190_{34}$	·98	$·8365_{24}$	·38	$·9162_{15}$
·19	$·5753_{40}$	·59	$·7224_{33}$	·99	$·8389_{24}$	·39	$·9177_{15}$
0·20	$0·5793_{39}$	0·60	$0·7257_{34}$	1·00	$0·8413_{25}$	1·40	$0·9192_{15}$
·21	$·5832_{39}$	·61	$·7291_{33}$	·01	$·8438_{23}$	·41	$·9207_{15}$
·22	$·5871_{39}$	·62	$·7324_{33}$	·02	$·8461_{24}$	·42	$·9222_{14}$
·23	$·5910_{38}$	·63	$·7357_{32}$	·03	$·8485_{23}$	·43	$·9236_{15}$
·24	$·5948_{39}$	·64	$·7389_{33}$	·04	$·8508_{23}$	·44	$·9251_{14}$
0·25	$0·5987_{39}$	0·65	$0·7422_{32}$	1·05	$0·8531_{23}$	1·45	$0·9265_{14}$
·26	$·6026_{38}$	·66	$·7454_{32}$	·06	$·8554_{23}$	·46	$·9279_{13}$
·27	$·6064_{39}$	·67	$·7486_{31}$	·07	$·8577_{22}$	·47	$·9292_{14}$
·28	$·6103_{38}$	·68	$·7517_{32}$	·08	$·8599_{22}$	·48	$·9306_{13}$
·29	$·6141_{38}$	·69	$·7549_{31}$	·09	$·8621_{22}$	·49	$·9319_{13}$
0·30	$0·6179_{38}$	0·70	$0·7580_{31}$	1·10	$0·8643_{22}$	1·50	$0·9332_{13}$
·31	$·6217_{38}$	·71	$·7611_{31}$	·11	$·8665_{21}$	·51	$·9345_{12}$
·32	$·6255_{38}$	·72	$·7642_{31}$	·12	$·8686_{22}$	·52	$·9357_{13}$
·33	$·6293_{38}$	·73	$·7673_{31}$	·13	$·8708_{21}$	·53	$·9370_{12}$
·34	$·6331_{37}$	·74	$·7704_{30}$	·14	$·8729_{20}$	·54	$·9382_{12}$
0·35	$0·6368_{38}$	0·75	$0·7734_{30}$	1·15	$0·8749_{21}$	1·55	$0·9394_{12}$
·36	$·6406_{37}$	·76	$·7764_{30}$	·16	$·8770_{20}$	·56	$·9406_{12}$
·37	$·6443_{37}$	·77	$·7794_{29}$	·17	$·8790_{20}$	·57	$·9418_{11}$
·38	$·6480_{37}$	·78	$·7823_{29}$	·18	$·8810_{20}$	·58	$·9429_{12}$
·39	$·6517_{37}$	·79	$·7852_{29}$	·19	$·8830_{19}$	·59	$·9441_{11}$

z	$\Phi(z)$	z	$\Phi(z)$	z	$\Phi(z)$	z	$\Phi(z)$
1·60	0.9452_{11}	2·00	0.97725_{53}	2·40	0.99180_{22}	2·75	0.99702_{9}
·61	$.9463_{11}$	·01	$.97778_{53}$	·41	$.99202_{22}$	·76	$.99711_{9}$
·62	$.9474_{10}$	·02	$.97831_{51}$	·42	$.99224_{21}$	·77	$.99720_{8}$
·63	$.9484_{11}$	·03	$.97882_{50}$	·43	$.99245_{21}$	·78	$.99728_{8}$
·64	$.9495_{10}$	·04	$.97932_{50}$	·44	$.99266_{20}$	·79	$.99736_{8}$
1·65	0.9505_{10}	2·05	0.97982_{48}	2·45	0.99286_{19}	2·80	0.99744_{8}
·66	$.9515_{10}$	·06	$.98030_{47}$	·46	$.99305_{19}$	·81	$.99752_{8}$
·67	$.9525_{10}$	·07	$.98077_{47}$	·47	$.99324_{19}$	·82	$.99760_{7}$
·68	$.9535_{10}$	·08	$.98124_{45}$	·48	$.99343_{18}$	·83	$.99767_{7}$
·69	$.9545_{9}$	·09	$.98169_{45}$	·49	$.99361_{18}$	·84	$.99774_{7}$
1·70	0.9554_{10}	2·10	0.98214_{43}	2·50	0.99379_{17}	2·85	0.99781_{7}
·71	$.9564_{9}$	·11	$.98257_{43}$	·51	$.99396_{17}$	·86	$.99788_{7}$
·72	$.9573_{9}$	·12	$.98300_{41}$	·52	$.99413_{17}$	·87	$.99795_{6}$
·73	$.9582_{9}$	·13	$.98341_{41}$	·53	$.99430_{16}$	·88	$.99801_{6}$
·74	$.9591_{8}$	·14	$.98382_{40}$	·54	$.99446_{15}$	·89	$.99807_{6}$
1·75	0.9599_{9}	2·15	0.98422_{39}	2·55	0.99461_{16}	2·90	0.99813_{6}
·76	$.9608_{8}$	·16	$.98461_{39}$	·56	$.99477_{15}$	·91	$.99819_{6}$
·77	$.9616_{9}$	·17	$.98500_{37}$	·57	$.99492_{14}$	·92	$.99825_{6}$
·78	$.9625_{8}$	·18	$.98537_{37}$	·58	$.99506_{14}$	·93	$.99831_{5}$
·79	$.9633_{8}$	·19	$.98574_{36}$	·59	$.99520_{14}$	·94	$.99836_{5}$
1·80	0.9641_{8}	2·20	0.98610_{35}	2·60	0.99534_{13}	2·95	0.99841_{5}
·81	$.9649_{7}$	·21	$.98645_{34}$	·61	$.99547_{13}$	·96	$.99846_{5}$
·82	$.9656_{8}$	·22	$.98679_{34}$	·62	$.99560_{13}$	·97	$.99851_{5}$
·83	$.9664_{7}$	·23	$.98713_{32}$	·63	$.99573_{12}$	·98	$.99856_{5}$
·84	$.9671_{7}$	·24	$.98745_{33}$	·64	$.99585_{13}$	·99	$.99861_{4}$
1·85	0.9678_{8}	2·25	0.98778_{31}	2·65	0.99598_{11}	3·0	0.99865_{38}
·86	$.9686_{7}$	·26	$.98809_{31}$	·66	$.99609_{12}$	3·1	$.99903_{28}$
·87	$.9693_{6}$	·27	$.98840_{30}$	·67	$.99621_{11}$	3·2	$.99931_{21}$
·88	$.9699_{7}$	·28	$.98870_{29}$	·68	$.99632_{11}$	3·3	$.99952_{14}$
·89	$.9706_{7}$	·29	$.98899_{29}$	·69	$.99643_{10}$	3·4	$.99966_{11}$
1·90	0.9713_{6}	2·30	0.98928_{28}	2·70	0.99653_{11}	3·5	0.99977_{7}
·91	$.9719_{7}$	·31	$.98956_{27}$	·71	$.99664_{10}$	3·6	$.99984_{5}$
·92	$.9726_{6}$	·32	$.98983_{27}$	·72	$.99674_{9}$	3·7	$.99989_{4}$
·93	$.9732_{6}$	·33	$.99010_{26}$	·73	$.99683_{10}$	3·8	$.99993_{2}$
·94	$.9738_{6}$	·34	$.99036_{25}$	·74	$.99693_{9}$	3·9	$.99995_{2}$
1·95	0.9744_{6}	2·35	0.99061_{25}				
·96	$.9750_{6}$	·36	$.99086_{25}$				
·97	$.9756_{5}$	·37	$.99111_{23}$				
·98	$.9761_{6}$	·38	$.99134_{24}$				
·99	$.9767_{5}$	·39	$.99158_{22}$				

$N(0, 1)$

$\Phi(z)$

0   z

Table 2

PERCENTAGE POINTS OF THE NORMAL DISTRIBUTION

P	10	5	2	1	0·2	0·1
x	1·64	1·96	2·33	2·58	3·09	3·29

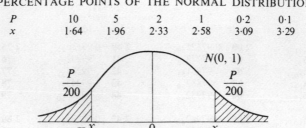

The table gives the value x so that $P(|z| \geqslant x) = P/100$ where z is $N(0, 1)$. (For a one-tailed test it is therefore necessary to look under the value $2P$ where P is the required significance level.)

Table 3

PERCENTAGE POINTS OF THE t-DISTRIBUTION

P	10	5	2	1	0·2	0·1
$v = 1$	6·31	12·71	31·82	63·66	318·3	636·6
2	2·92	4·30	6·96	9·92	22·33	31·60
3	2·35	3·18	4·54	5·84	10·21	12·92
4	2·13	2·78	3·75	4·60	7·17	8·61
5	2·02	2·57	3·36	4·03	5·89	6·87
6	1·94	2·45	3·14	3·71	5·21	5·96
7	1·89	2·36	3·00	3·50	4·79	5·41
8	1·86	2·31	2·90	3·36	4·50	5·04
9	1·83	2·26	2·82	3·25	4·30	4·78
10	1·81	2·23	2·76	3·17	4·14	4·59
12	1·78	2·18	2·68	3·05	3·93	4·32
15	1·75	2·13	2·60	2·95	3·73	4·07
20	1·72	2·09	2·53	2·85	3·55	3·85
24	1·71	2·06	2·49	2·80	3·47	3·75
30	1·70	2·04	2·46	2·75	3·39	3·65
40	1·68	2·02	2·42	2·70	3·31	3·55
60	1·67	2·00	2·39	2·66	3·23	3·46
120	1·66	1·98	2·36	2·62	3·16	3·37
∞	1·64	1·96	2·33	2·58	3·09	3·29

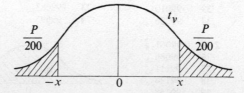

The table gives the value x so that $P(|t_v| \geqslant x) = P/100$ where t_v has the t-distribution with v degrees of freedom.

Table 4

PERCENTAGE POINTS OF THE χ^2-DISTRIBUTION

P	99	95	10	5	1	0·1
$v = 1$	0.000157	0·00393	2·71	3·84	6·63	10·83
2	0·0201	0·103	4·61	5·99	9·21	13·81
3	0·115	0·352	6·25	7·81	11·34	16·27
4	0·297	0·711	7·78	9·49	13·28	18·47
5	0·554	1·15	9·24	11·07	15·09	20·52
6	0·872	1·64	10·64	12·59	16·81	22·46
7	1·24	2·17	12·02	14·07	18·48	24·32
8	1·65	2·73	13·36	15·51	20·09	26·12
9	2·09	3·33	14·68	16·92	21·67	27·88
10	2·56	3·94	15·99	18·31	23·21	29·59
12	3·57	5·23	18·55	21·03	26·22	32·91
14	4·66	6·57	21·06	23·68	29·14	36·12
16	5·81	7·96	23·54	26·30	32·00	39·25
18	7·01	9·39	25·99	28·87	34·81	42·31
20	8·26	10·85	28·41	31·41	37·57	45·31
25	11·52	14·61	34·38	37·65	44·31	52·62
30	14·95	18·49	40·26	43·77	50·89	59·70
40	22·16	26·51	51·81	55·76	63·69	73·40
50	29·71	34·76	63·17	67·50	76·15	86·66
60	37·48	43·19	74·40	79·08	88·38	99·61
70	45·44	51·74	85·53	90·53	100·4	112·3
80	53·54	60·39	96·58	101·9	112·3	124·8
90	61·75	69·13	107·6	113·1	124·1	137·2
100	70·06	77·93	118·5	124·3	135·8	149·4

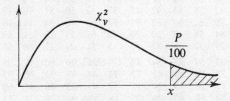

The table gives the value x so that $P(\chi_v^2 \geqslant x) = P/100$ where χ_v^2 has the χ^2-distribution with v degrees of freedom.

Table 5 TABLE OF RANDOM NUMBERS

09	33	37	47	56	24	88	58	95	22	00	41	04	14	57	95	54	20	54	66
18	33	71	44	25	74	04	98	19	78	50	47	41	45	17	96	50	87	35	44
75	40	23	87	41	72	28	20	33	30	45	76	11 -	10	08	16	09	21	51	60
72	71	20	29	58	19	85	17	25	84	03	57	06	61	98	29	26	00	13	01
73	06	34	58	39	03	61	74	78	74	24	37	75	49	32	85	18	19	47	13
92	17	02	18	77	52	81	61	55	71	72	69	21	41	88	74	71	42	97	95
50	46	43	54	39	34	49	46	60	02	59	90	08	86	36	18	56	11	19	77
47	52	78	86	12	66	10	55	73	75	35	64	23	22	47	59	82	01	78	22
96	23	65	45	24	25	10	13	33	57	32	70	56	52	92	89	02	43	73	01
05	43	27	33	01	44	46	69	38	64	32	03	14	74	71	56	38	80	92	35
45	32	58	59	44	01	90	22	83	22	51	43	09	22	92	07	14	24	98	59
93	49	37	18	03	79	05	05	69	56	94	75	60	58	32	92	99	64	73	86
14	19	12	52	64	33	08	99	68	50	44	10	59	53	27	79	71	12	90	65
47	11	66	74	71	55	87	77	87	27	59	93	89	85	90	66	23	22	75	41
45	71	82	13	74	27	83	15	33	23	47	28	77	67	59	89	67	55	64	54
36	71	69	93	77	32	19	50	85	59	72	84	05	98	72	88	73	88	44	00
69	12	63	23	53	76	53	07	99	40	91	76	54	62	11	61	24	60	93	83
54	67	32	10	13	00	08	36	44	57	62	19	64	07	11	98	29	27	78	44
59	36	34	90	14	93	87	07	26	56	57	53	54	21	65	47	78	31	17	22
06	07	21	10	97	69	72	19	71	37	79	20	21	18	22	06	06	21	10	98
71	77	26	80	49	94	52	32	21	81	20	27	48	95	41	47	49	57	59	14
59	03	09	00	36	44	15	08	56	72	29	63	83	52	06	54	58	29	40	02
76	97	77	91	48	11	90	45	13	93	44	03	68	00	35	75	80	75	71	29
60	25	34	73	04	62	58	53	73	79	82	43	73	32	32	58	71	07	77	42
13	62	01	16	55	29	13	42	28	57	76	22	06	88	80	54	47	69	82	95
96	44	52	78	36	09	60	99	12	47	15	77	57	64	74	96	91	70	94	74
70	94	07	18	90	32	18	85	42	77	37	32	50	51	70	19	59	99	02	91
25	17	41	08	04	76	62	60	33	92	98	62	70	83	11	16	08	48	59	28
02	18	87	94	89	84	38	63	20	20	48	65	95	23	11	69	58	02	86	71
68	53	34	75	85	25	32	16	37	33	13	51	09	21	86	66	83	23	48	68
43	98	43	39	22	03	67	77	54	88	39	20	90	60	17	08	44	87	60	48
12	64	63	47	13	23	86	89	81	72	22	44	97	93	05	93	69	94	93	28
10	99	07	19	83	11	76	13	34	11	57	43	32	46	46	68	23	97	51	95
03	86	04	69	57	79	10	63	98	25	18	88	96	78	31	18	48	83	58	72
10	34	55	33	92	42	44	35	25	00	21	24	20	83	51	21	54	61	02	02
21	70	23	20	60	02	70	51	05	57	68	48	74	58	59	14	34	79	60	51
12	55	40	78	43	45	70	79	08	68	19	73	02	19	95	42	35	10	77	22
07	55	26	44	94	38	12	26	42	94	82	91	83	68	06	09	38	04	79	00
78	88	43	17	20	15	36	83	15	57	19	62	10	23	24	75	99	81	71	78
64	40	66	53	63	30	67	33	40	73	93	37	15	88	67	37	47	09	82	03
82	93	47	20	23	17	38	84	16	57	19	61	08	21	21	72	95	77	66	72
58	33	59	45	54	21	57	48	54	99	45	25	53	38	67	00	48	98	09	08
04	60	98	05	31	88	36	03	98	82	93	29	51	71	48	20	96	38	81	25
29	30	39	41	97	44	92	27	09	76	37	79	50	60	97	20	40	18	66	56
49	69	77	83	73	09	00	09	44	81	57	69	04	23	36	05	55	73	20	21
39	83	27	07	34	69	47	67	15	52	14	10	16	41	60	07	82	70	32	91
08	95	37	94	91	88	20	00	87	93	90	17	85	54	36	03	66	87	01	09
73	04	75	24	61	62	39	64	49	54	16	44	26	21	54	35	51	48	39	09
20	06	79	13	42	66	45	03	51	49	33	05	26	32	33	06	11	86	40	50
59	77	90	58	94	57	58	59	69	51	14	20	80	53	65	00	19	34	04	56

Table 6
TABLE OF e^{-x} FOR VALUES FROM 0·0 TO 9·9

e^{-x}	0·0	0·1	0·2	0·3	0·4
0	1·0000	0·9048	0·8187	0·7408	0·6703
1	0·3679	0·3329	0·3012	0·2725	0·2466
2	0·1353	0·1225	0·1108	0·1003	0·0907
3	0·0498	0·0450	0·0408	0·0369	0·0334
4	0·0183	0·0166	0·0150	0·0136	0·0123
5	0·0067	0·0061	0·0055	0·0050	0·0045
6	0·0025	0·0022	0·0020	0·0018	0·0017
7	0·0009	0·0008	0·0007	0·0007	0·0006
8	0·0003	0·0003	0·0003	0·0002	0·0002
9	0·0001	0·0001	0·0001	0·0001	0·0001

e^{-x}	0·5	0·6	0·7	0·8	0·9
0	0·6065	0·5488	0·4966	0·4493	0·4066
1	0·2231	0·2019	0·1827	0·1653	0·1496
2	0·0821	0·0743	0·0672	0·0608	0·0550
3	0·0302	0·0273	0·0247	0·0224	0·0202
4	0·0111	0·0101	0·0091	0·0082	0·0074
5	0·0041	0·0037	0·0033	0·0030	0·0027
6	0·0015	0·0014	0·0012	0·0011	0·0010
7	0·0006	0·0005	0·0005	0·0004	0·0004
8	0·0002	0·0002	0·0002	0·0002	0·0001
9	0·0001	0·0001	0·0001	0·0001	0·0001

Table 7
TABLE OF $\log_{10} n!$ FOR VALUES FROM 1 TO 100

n	$\log n!$	n	$\log n!$	n	$\log n!$	n	$\log n!$	n	$\log n!$
1	0·0000	21	19·7083	41	49·5244	61	83·7055	81	120·7632
2	0·3010	22	21·0508	42	51·1477	62	85·4979	82	122·6770
3	0·7782	23	22·4125	43	52·7811	63	87·2972	83	124·5961
4	1·3802	24	23·7927	44	54·4246	64	89·1034	84	126·5204
5	2·0792	25	25·1906	45	56·0778	65	90·9163	85	128·4498
6	2·8573	26	26·6056	46	57·7406	66	92·7359	86	130·3843
7	3·7024	27	28·0370	47	59·4127	67	94·5619	87	132·3238
8	4·6055	28	29·4841	48	61·0939	68	96·3945	88	134·2683
9	5·5598	29	30·9465	49	62·7841	69	98·2333	89	136·2177
10	6·5598	30	32·4237	50	64·4831	70	100·0784	90	138·1719
11	7·6012	31	33·9150	51	66·1906	71	101·9297	91	140·1310
12	8·6803	32	35·4202	52	67·9066	72	103·7870	92	142·0948
13	9·7943	33	36·9387	53	69·6309	73	105·6503	93	144·0632
14	10·9404	34	38·4702	54	71·3633	74	107·5196	94	146·0364
15	12·1165	35	40·0142	55	73·1037	75	109·3946	95	148·0141
16	13·3206	36	41·5705	56	74·8519	76	111·2754	96	149·9964
17	14·5511	37	43·1387	57	76·6077	77	113·1619	97	151·9831
18	15·8063	38	44·7185	58	78·3712	78	115·0540	98	153·9744
19	17·0851	39	46·3096	59	80·1420	79	116·9516	99	155·9700
20	18·3861	40	47·9116	60	81·9202	80	118·8547	100	157·9700

Answers

Answers

Exercise 2A, p. 13.

1. (a) $\frac{5}{8}$; (b) $\frac{1}{2}$; (c) $\frac{5}{8}$; (d) $\frac{1}{4}$. **2.** (a) $\frac{1}{2}$; (b) $\frac{1}{4}$; (c) $\frac{1}{12}$; (d) $\frac{1}{6}$. **3.** (ii) (a) $\frac{5}{24}$; (b) $\frac{5}{8}$; (c) $\frac{1}{8}$; (iii) $\frac{4}{7}$. **4.** 0·572. **5.** (i) $\frac{1}{4}$; (ii) $\frac{7}{12}$; (iii) $\frac{1}{3}$. **6.** $2/n$. **7.** (i) $\frac{1}{720}$; (ii) $\frac{1}{120}$; (iii) $\frac{1}{6}$. **8.** (a) 6249; (b) 113; (a) $\frac{149}{6249}$; (b) $\frac{47}{113}$. **9.** (a) $7! = 5040$; (b) $7!/2! = 2520$; (c) $7!/(3!2!) = 420$; $n!/(p_1!p_2!\ldots)$ **10.** $8!/2! = 20160$; $\frac{1}{4}$. **11.** (i) $\frac{5}{9}$; (ii) $\frac{7}{12}$. **12.** 50 400; 35 700; 8500; $\frac{1}{504}$; $\frac{1}{357}$; 0. **13.** (a) $\frac{2}{5}$; (b) $\frac{1}{10}$; (c) $\frac{1}{10}$. **14.** 1260; (i) $\frac{1}{36}$; (ii) $\frac{1}{420}$; (iii) $\frac{1}{105}$; (iv) $\frac{5}{12}$. **15.** $\frac{1}{3}$; $\frac{2}{9}$; $\frac{16}{81}$.

Exercise 2B, p. 19.

1. (i) $\frac{5}{91}$; (ii) $\frac{6}{91}$. **2.** 92; (i) $\frac{7}{23}$; (ii) $\frac{9}{23}$. **3.** $2^{13}/\binom{48}{6} \simeq 7 \times 10^{-4}$. **6.** $12!/(4!)^3 = 34650$; (i) $\frac{1}{11}$; (ii) $\frac{3}{11}$. **8.** (a) $\frac{16}{21}$; (b) $\frac{11}{21}$. **9.** (a) 0·2; (b) 0·9. **11.** (a) $\frac{2}{15}$; (b) $\frac{13}{15}$; (c) $\frac{2}{5}$; (d) $\frac{4}{5}$. **12.** (i) 0·85; (ii) 0·10; (iii) 0·20. **13.** (i) $\frac{1}{15}$; (ii) $\frac{14}{45}$; (iii) $\frac{31}{45}$. **14.** $0·9 \times 10^{-7}$. **15.** (i) 0·0023; (ii) 0·10; (iii) 0·35.

Exercise 3A, p. 28.

1. (i) $\frac{2}{15}$; (ii) $\frac{4}{15}$; (iii) $\frac{3}{5}$. **2.** (b) 0·067. **4.** (a) $\frac{1}{20}$; (b) $\frac{7}{15}$; (c) $\frac{7}{20}$. **7.** (i) $\frac{1}{15}$; (ii) $\frac{31}{90}$; $\frac{3}{10}$. **8.** (i) $\frac{2}{15}$; (ii) $\frac{2}{3}$; (iii) $\frac{17}{150}$; (iv) $\frac{19}{100}$. **9.** (i) $\frac{3}{5}$; (ii) $\frac{43}{55}$; (iii) $\frac{7}{11}$; (iv) $\frac{14}{25}$. **10.** (i) $\frac{27}{100}$; (ii) $\frac{69}{100}$; (iii) $\frac{9}{23}$; (iv) $\frac{1}{20}$; (v) $\frac{3}{31}$. **11.** (i) $\frac{8}{45}$; (ii) $\frac{2}{5}$. **13.** (ii) $\frac{1}{5}$. **14.** (ii) $\frac{1}{8}$; (iii) $\frac{1}{2}$; (iv) $\frac{1}{4}$. **16.** (i) $\frac{1}{20}$, $\frac{1}{2}$, $\frac{1}{19}$, $\frac{10}{19}$; (ii) $\frac{1}{4}$, $\frac{4}{5}$, $\frac{79}{304}$, $\frac{79}{95}$. **17.** (a) $\frac{1}{12}$; (b) $\frac{5}{648}$; (c) $\frac{31}{216}$. **18.** $1 - (\frac{2 \cdot 47 \cdot 46}{13 \cdot 17 \cdot 25})^2 \simeq 0·387$. **19.** (a) 0·35; (b) 0·235. **20.** $\frac{1}{2}$.

Exercise 3B, p. 35.

1. $\frac{1}{7}$. **2.** $\frac{40}{73}$, $\frac{24}{73}$, $\frac{9}{73}$. **3.** $\frac{3}{5}$. **4.** (i) $\frac{1}{3}$, $\frac{1}{4}$, $\frac{5}{12}$; (ii) $\frac{1}{9}$. **5.**

	N	T	R	TR
H	$\frac{24}{29}$	$\frac{3}{29}$	$\frac{28}{435}$	$\frac{2}{435}$
W	0	$\frac{36}{47}$	$\frac{8}{47}$	$\frac{3}{47}$
S	0	$\frac{1}{2}$	$\frac{2}{9}$	$\frac{5}{18}$

6. $\frac{143}{250}$; $\frac{45}{143}$. **7.** $\frac{5}{1116}$. **8.** $\frac{35}{64}$; $\frac{4}{7}$, $\frac{12}{35}$. **9.** (i) $\frac{5}{8}$; (ii) $\frac{77}{128}$. **10.** $\frac{1}{3}$, $\frac{11}{24}$.

Revision Examples I, p. 37.

1. $\frac{2}{3}$, $\frac{1}{3}$. **2.** 57. **3.** $\frac{2}{7}$; $3p^2(1-p)$. **4.** 0·30, 0·42. **5.** (i) 46%; (ii) 16·1%; (iii) 13%; (iv) 82·9%. **6.** (i) $\frac{55}{96}$; (ii) $\frac{55}{144}$. **7.** 1, 2, 9. **9.** (i) $\frac{9}{100}$; (ii) $\frac{111}{1000}$; (iii) $\frac{37}{110}$; (iv) small; (v) no. **10.** $\frac{27}{59}$; $\frac{729}{1241}$. **13.** $\frac{14}{45}$.

14. $[1-p^6(1+6q+21q^2+56q^3+126q^4)-q^6(1+6p+21p^2+56p^3+126p^4)](2pq)^{n-5}$
(p^2+q^2). **15.** (i) $\frac{3}{20}$; (ii) $\frac{1}{6}$; (iii) $\frac{59}{60}$. **16.** $\frac{1}{5}$;

	0	1	2	3	
(a)	$\frac{64}{125}$	$\frac{48}{125}$	$\frac{12}{125}$	$\frac{1}{125}$	$\frac{12}{13}$.
(b)	$\frac{1}{125}$	$\frac{12}{125}$	$\frac{48}{125}$	$\frac{64}{125}$	

17. $\frac{78}{115}$; $\frac{5}{17}$. **18.** (a) $\frac{5}{7}$; (b) $\frac{3}{5}$; (c) $\frac{12}{125}$; $\frac{4}{5}$.

19. $p^3[p+(r-p)q^3]$; $p^3[p+\dfrac{(r-p)}{15}(1+2q+4q^2+8q^3)]$. **20.** (i) 0.52; (ii) 0.58; 55%.

Exercise 4, p. 55.

3. 80.8. **4.** (i) 34.78. **5.** 42. **6.** (a) 5.5; (b) 0.55; (c) 141 m. **9.** (a) 15.2, 5–9; (b) 25.6, 15–19.

Exercise 5, p. 64.

1. 143. **2.** 4.6; 4. **3.** (b) (i) 26.0; (ii) 25.4; (c) (i) 27; (ii) 6; (iii) 26.
4. (ii) 65; (iii) 78%. **5.** (a) 89, 88; (b) 39, 38.

Exercise 6A, p. 70.

1. 12; 11; 13. **2.** 1.5; 143; 144. **3.** (a) 179.7 cm; (b) 176.2 cm, 183.5 cm;
(c) 285. **4.** (a) £1150; (b) £1265, £1010; (c) 655. **5.** (a) 49, 45.5; (b) 25, 20.

Exercise 6B, p. 79.

1. A 15, 2.74; B 15, 1.35. **2.** 1.11. **3.** £160. **4.** (i) 100.45; (ii) 10.03;
(iii) 13.44. **5.** 6.20; 2.42; 1, $11\frac{1}{2}$—13; $\frac{2}{25}$. **6.** $\bar{x}$; M. **7.** 11 years 4 months;
6.62 months. **8.** $nS_1^2/(n+1)$. **9.** 45.7, 1.60; 46.0, 1.72. **10.** $(m\bar{x}+n\bar{y})/(m+n)$.
11. 179.8; 4.91. **12.** 56.90; 26.10; 45.9%.

Revision Examples II, p. 81.

2. 57.2. **3.** £67; £58; £75. **4.** (i) $\frac{1}{3}(a+b+c)$; (ii) b; (iii) $\frac{1}{3}(c-a)$;
(iv) $\sqrt{[\frac{1}{3}(a^2+b^2+c^2)-\frac{1}{9}(a+b+c)^2]}$; $\sqrt{[\frac{1}{4}(a^2+b^2+c^2+d^2)-\frac{1}{16}(a+b+c+d)^2]}$
5. 48.7; 21.50; 72; 48; 25. **6.** 3.87; 1.92. **7.** (i) 7 minutes 52 seconds;
(ii) 63%; 7 minutes 40.1 seconds; 17.4 seconds. **8.** 9; 4; 25, 35, 45, 65, 80.
9. 50.5; 9.91. **10.** $\simeq 130$ (depending on grouping). **11.** 18.43; 6.26; 17.9;
13.7; 22.3.

Exercise 7A, p. 90.

2. 52.5p. **3.** 8, 5; 9, 5. **4.** $E(x)+a$; $\text{Var}(x)$. **5.** 6.75p. **6.** 1.82 days.
7. 104 m^2; 35.3 m^2. **8.** £$\frac{25}{7}$, $\frac{173}{49}$; £25, 9 or 33. **9.** £3; £9;

$aE(x)$; $a^2\text{Var}(x)$. **10.** $P(x)=\dfrac{x-1}{36}$, $2 \leqslant x \leqslant 7$, $\dfrac{13-x}{36}$, $8 \leqslant x \leqslant 12$; 55.6p.

11. 7; $\frac{35}{6}$. **12.** 4.5; $\frac{33}{4}$. **14.** $\frac{8}{3}$; $\frac{20}{9}$. **15.** (a)$\frac{1}{3}$; (b) $\frac{5}{12}$; (c) $\frac{1}{12}$; $\frac{4}{27}$.
16. $P(0)=\frac{125}{216}$, $P(6)=\frac{25}{216}$, $P(x)=(x-2)/216$, $7 \leqslant x \leqslant 12$, $P(x)=(19-x)/216$,
$13 \leqslant x \leqslant 18$. **17.** $\frac{8}{5}$; $\frac{16}{25}$. **18.** $-\frac{19}{18}$. **19.** $p(2p^2+p+1)$. **20.** (i) 8.75;
(ii) 5.25.

Exercise 7B, p. 98.

1. $P(r) = \frac{1}{6}(\frac{5}{6})^{r-1}$, $r = 1, 2 \ldots$; (i) 1; (ii) 6; $F(r) = 1 - (\frac{5}{6})^r$, $r = 1, 2 \ldots$; 4. **2.** (a) $\frac{1}{16}$; (b) $= \frac{1}{32}$; 2^{-19}; infinite. **3.** $p_1 = p^2$; $p_2 = qp^2$. **4.**

r	0	1	2	3	4	5
$F(r)$	$\frac{1}{6}$	$\frac{1}{2}$	$\frac{1}{2}$	$\frac{3}{4}$	$\frac{11}{12}$	1

$\frac{13}{6}$; $\frac{95}{36}$. **5.** $\frac{8}{27}$; $\frac{16}{81}$; $\frac{8}{3}$; $\frac{8}{9}$. **6.** $\frac{1}{3}$, $\frac{1}{6}$, $\frac{1}{2}$. **7.** $t/(3-2t)$; 3; 6. **8.** t/p. **9.** $\frac{4}{9}$; $\frac{44}{9}$.

Exercise 8A, p. 106.

1. (i) $\frac{5}{16}$; (ii) $\frac{13}{16}$; (iii) $\frac{1}{16}$. **2.** 577/117 649. **3.** (a) $\frac{16}{81}$; (b) $\frac{32}{81}$; (c) $\frac{1}{9}$. **6.** $1 - 104.6^5/7^7 \simeq 0.018$. **7.** (a) $\frac{5}{16}$; (b) $\frac{1}{32}$. **8.** (a) $2(\frac{5}{7})^6 \simeq 0.27$; (b) $(\frac{5}{7})^7 \simeq 0.095$; (c) $1 - 2(\frac{5}{7})^6 - (\frac{5}{7})^7 \simeq 0.64$. **9.** 0.384; yes, $P \simeq 0.97$.

10. (c); $\frac{37}{128}$. **11.** $P(x) = \frac{n-x+1}{x} \cdot \frac{p}{q} P(x-1)$; 0.444, 0.391, 0.138, 0.024, 0.002, 0.000. **12.** (a) 3.5; (b) 3; 3. **13.** (a) $(n+1)p = m + \alpha$, $0 < \alpha < 1$, mode $= m$, (b) $(n+1)p$ and $(n+1)p-1$ are the two modes. **15.** (a) 0.00037; (b) 0.0035. **16.** £$\frac{3}{128}$. **17.** 0.71; 0.33. **18.** $(1-q^3)^4[10q^6 + 4q^3 + 1]$. **19.** 36; 48/$\sqrt{n}$. **20.** £1.76, 0.468.

Exercise 8B, p. 114.

1. 2.9. **2.** (a) 0.050; (b) 0.050; (c) 0.353. **3.** 0.0018. **4.** 0.801. **5.** 0.030, 0.106, 0.185, 0.216, 0.189, 0.132, 0.077; 3; 4.25. **6.** 8. **7.** 2.53; (a) 0.28; (b) 0.35. **8.** $e^{\mu(t-1)}$. **9.** $\mu = m + a$, $0 < a < 1$, mode $= m$; bimodal if $\mu = m$, modes m and $m-1$. **11.** 0.007; 10; 0.125. **12.** 339, 504, 375, 186, 69, 21. **13.** $\mathscr{P}(1.68)$. **14.** 1.1. **15.** (a) $\log(1/p)$;

$$(b) \quad p\left[\frac{(\ln(\frac{1}{p}))^{2x-1}}{(2x-1)!} + \frac{(\ln(\frac{1}{p}))^{2x}}{(2x)!}\right] x \geqslant 1; \ (c) \ (1-p^2)/2.$$

Exercise 9A, p. 127.

1. (i) $f(x) = 2x$, $0 \leqslant x \leqslant 1$;

$$(ii) \quad F(x) = \begin{cases} 0 & x \leqslant 0 \\ x^2 & 0 < x \leqslant 1 \\ 1 & x > 1 \end{cases};$$

(iii) $\frac{2}{3}$, $1/\sqrt{2}$; (iv) $\frac{1}{4}$.

2. $f(x) = \frac{1}{b-a}$, $a \leqslant x \leqslant b$; $\qquad F(x) = \begin{cases} 0 & x \leqslant a \\ \dfrac{x-a}{b-a} & a < x \leqslant b \\ 1 & x > b \end{cases}$

$\frac{1}{2}(b+a)$; $\frac{1}{12}(b-a)^2$.

3. $f(x) = \begin{cases} 1+x & -1 \leqslant x \leqslant 0 \\ 1-x & 0 < x \leqslant 1 \end{cases}$; $\qquad F(x) = \begin{cases} 0 & x \leqslant -1 \\ x + \frac{1}{2}(1+x^2) & -1 < x \leqslant 0 \\ x + \frac{1}{2}(1-x^2) & 0 < x \leqslant 1 \\ 1 & x > 1 \end{cases}$

$\frac{1}{6}$. **4.** $\frac{1}{9}\sqrt{1850} \simeq 4.8$. **5.** $1 - 10^{-2/5} \simeq 0.60$ thousand litres. **7.** $\pi/10$; $\frac{5}{2}$; $\frac{5}{3}$; $\frac{10}{3}$. **8.** $f(x) = \frac{1}{9}x^2$, $0 \leqslant x \leqslant 3$; $\frac{27}{80}$. **9.** 2 log 2; $2 - (2 \log 2)^2$; $\frac{1}{3}$; $\frac{4}{3}$; $\frac{8}{7}$; $\frac{8}{5}$; $\frac{16}{35}$. **10.** $\frac{3}{64}$; 4; 0.138; 2.14.

11. $4; f(x) = \begin{cases} 2x/5 & 0 \leqslant x \leqslant 1 \\ \frac{2}{5} & 1 < x \leqslant 2; \\ \frac{1}{5}(4-x) & 2 < x \leqslant 4 \end{cases}$ $\frac{3}{8}$.

12. $\frac{97}{60}; \frac{77}{160}$. **13.** $-1, 8, -6, \frac{2}{3}$. **14.** 2·51 pm; $\frac{1}{6}$. **15.** $\frac{11}{36}$. **16.** $a\sqrt{(\frac{8}{3} - \frac{1}{4}\pi^2)}$

17. $\frac{15}{1024}; \frac{16}{7}, \frac{24}{49}$. **19.** $\frac{1}{6}a^2$. **20.** $\frac{3}{32}; 2; 2/\sqrt{5}; 2; 1 - \frac{7}{25}\sqrt{5}$.

Exercise 9B, p. 130.

1. (a) $1 - e^{-2/3} \simeq 0.49$; (b) $e^{-4/3} \simeq 0.26$; $30\,000 \log_e 2$.

2. $(1/\mu)e^{-t/\mu}$, $t \geqslant 0$; $e^{-2} \simeq 0.14$; $5/\log_e \frac{1000}{999} \simeq 5400$ hours.

3. $\frac{1}{\lambda^2}; \frac{\ln 3}{\lambda}$.

5. $3e^{-3x}$, $x \geqslant 0$; $e^{-1} \simeq 0.37$. **6.** $\frac{1}{3}(2e^{-\frac{1}{2}} - e^{-1} - e^{-2}) \simeq 0.24$; $\frac{10}{3}$ years; $\frac{8}{9}$ years^2.

7. $f(x) = \dfrac{1}{2\sqrt{x}}$, $0 \leqslant x \leqslant 1$; $\frac{1}{3}, \frac{4}{45}$.

8. $f(t) = 2/t^2$, $1 \leqslant t \leqslant 2$; $\frac{4}{21}$. **9.** $1 - (2/\pi)\tan^{-1}2$.

10. $f(x) = \dfrac{x}{2a\sqrt{(4a^2 - x^2)}}$, $0 \leqslant x \leqslant 2a$; $1 - \frac{1}{2}\sqrt{3}$.

11. $f(x) = \frac{1}{1800}x$, $0 \leqslant x \leqslant 60$; $40, 200$;

$f(x) = \dfrac{60 - x}{1800}$, $0 \leqslant x \leqslant 60$; $20, 200$.

Exercise 10A, p. 140.

1. (i) 0·875; (ii) 0·258; (iii) 0·375; (iv) 0·617; (v) 0·484. **2.** (i) 0·599; (ii) 0·290, (iii) 0·841; (iv) 0·052. **3.** (i) 0·345; (ii) 0·655. **4.** 0·0027. **6.** (i) 5%; (ii) 103·26; (iii) no. **7.** 46·67; $\frac{2}{3}(x + 20)$. **8.** 0·50; (a) 15·9%; (b) 6·7%. **10.** 114·10; 7·96. **11.** 0·16%, 16·4%. **12.** 90·1%. **13.** $N(a\mu + b, a^2\sigma^2)$. **14.** 13·4%; 1·63; 4·98. **16.** 20·6, 98·7, 188·0, 142·7, 43·0, 5·1. **17.** 0·196, 0·258, out of control. **18.** $110 + 20/\sqrt{(2\pi)} \simeq 117.98$g. **19.** 73·5. **20.** 4·971, 5, 5·029; 0·23.

Exercise 10B, p. 145.

1. 10, $\sqrt{5}$, 0·059. **2.** $B(20, 0.1)$. **3.** 0·02. **4.** 0·79. **5.** 2; 2·11. **6.** (i) 0·84; (ii) 0·71. **7.** 11.

Revision Examples III, p. 146.

1. $\frac{23}{35}; \frac{2}{3}; \frac{9}{5}$. **2.** $1 - 6e^{-5} + 15e^{-4} - 10e^{-3} \simeq 0.73$. **3.** 7·51 am.

4. $2 + \dfrac{1}{\lambda}$; $2 + \dfrac{\ln 2}{\lambda}$; $\dfrac{1}{\lambda}$. **5.** $A_L = 12.3$; $A_U = 13.5$.

6. 0·050, 0·149, 0·224, 0·224, 0·168, 0·101, 0·084; (i) £157·8; (ii) £168·1; (iii) £161·6.

7. $\dfrac{3\pi}{4} + \dfrac{1}{\pi}$; $\dfrac{4}{\pi}$. **8.** (i) 8·48; (ii) 5 min; 118; 4.

9. $F(x) = \dfrac{1}{1+e^{-\pi x/\sigma\sqrt{3}}}$; 0.860; 0.974; 0.996.

10. (i) 0.00031; (ii) 1.8, 1.26; (iii) 0.324. **11.** $\frac{3}{4}$; $F(t) = \frac{1}{2}t + \frac{1}{4}$, $0 \leqslant t < \frac{3}{2}$; $\frac{9}{16}$; $\frac{63}{256}$.

12. $(1-p)^{10}\{1 + 10p(1-p)^9\}$; 13.15.

13. $b = \dfrac{a^2}{1-a}$; 3; 5. **14.** $f(x) = 1+x$, $0 \leqslant x \leqslant \sqrt{3}-1$; $\frac{7}{64}$.

15. $120 - 50\Phi(2) - 25e^{-2}/\sqrt{(2\pi)} \simeq 70\text{p}$.

16. $\frac{4}{25}xe^{-2x/5}$; $\dfrac{5\sqrt{2}}{2}$. **17.** $M(t) = \dfrac{e^{3t}-e^t}{2t}$.

18. (i) $P(0) = P(4) = \frac{1}{70}, P(1) = P(3) = \frac{16}{70}$, $P(2) = \frac{36}{70}$; (ii) $\frac{16}{7}$.

(i) $P(0) = P(3) = \frac{1}{14}$, $P(1) = P(2) = \frac{3}{7}$; $\frac{3}{2}$; (ii) $\frac{15}{7}$.

19. (i) $(pe^{-\mu}+q)^6$; (ii) $6p(1-e^{-\mu})$; (iii) $p\mu$.

20. $\frac{2611}{900}$; $\dfrac{2611(30-n)}{30^3}$.

Exercise 11A, p. 158.

1. 0.087; 0.0073. **2.** $\bar{x}$; 2. **5.** $E(x_1) - E(x_2)$; $\text{Var}(x_1) + \text{Var}(x_2)$. **6.** 7; $\frac{35}{2}$.

7. p; pq. **8.** $\frac{9}{2}$; $33/4n$. **9.** (i) 4.02; (ii) 0.030; (iii) 0.071. **10.** 33. **11.** 0.975.

12. 2.08, 1.16, 3.12. **13.** (i) 5; (ii) 10; (iii) 5; (iv) 5; (v) 5. **14.** $\frac{5}{4}$, $\frac{5}{16}$;

$\dfrac{50}{51-i}$ $\dfrac{50(i-1)}{(51-i)^2}$. **16.** $a_1 = a_2 = \frac{1}{2}$; $a_1 = a_2 = \ldots = a_n = 1/n$; 5.0; 0.6.

17. $a = \frac{5}{9}$; $b = \frac{2}{9}$. **18.** 25; $\dfrac{24M(50-M)}{49}$. **19.** $\dfrac{\mu}{1-e^{-\mu}}$. **20.** $-p\left(1 + \dfrac{\ln p}{1-p}\right)$.

Exercise 11B, p. 162.

1. 60 min; 3.16 min; (a) 0.006; (b) 0.057; (c) 0.565. **2.** 0.106; 1242.8. **3.** 58%;

739 g. **4.** (a) 0.48; (b) 0.43; (c) 0.50. **5.** (i) 3.6%; (ii) 5.5%; 24.1%.

6. 0.84; 0.98.

For the remaining exercises the significance level is 5% unless otherwise stated.

Exercise 12A, p. 172.

1. $P(\geqslant 4 \text{ prefer A}) = 0.187$, hence not significant. **2.** 0.2; no, $p = 0.091$.

3. $p = 0.0625$, not significant; 1 set. **4.** (a) reject H_0 if $x < 0.316$; (b) reject H_0 if

$x < 0.224$ or $x > 1.776$; $p = 0.342$. **6.** $p = 0.24$, not significant. **7.** $z = 2.17$,

significant. **8.** (i) 0.201; (ii) 0.322; 0.137; $z = 1.77$, not significant; $z = 3.54$,

significant. **9.** $z = 1.36$, not significant; $z = 2.50$, not significant. **10.** $z = -3.33$,

significant. **11.** $z = 1.55$, not significant; 48, 65. **12.** $z = 2.77$, significant;

0.5455 ± 0.0425. **13.** 0.3 ± 0.042; 1.5 ± 0.212. **14.** 0.05; 0.006. **15.** $z = 2$,

significant.

Exercise 12B, p. 178.

1. $z = 2$, significant. **2.** $z = 2\cdot4$; 5%, 1%, not 0·5%. **3.** $z = 1\cdot725$, not significant at 5%, significant at 10%; $z = 1\cdot725$, significant at 5%. **4.** 0·0036; 5·008 $\pm$ 0·007; $z = 2\cdot2$; (i) significant; (ii) not significant; $z = 1\cdot27$, not significant—no further adjustment. **5.** $z = 2\cdot8$, significant; 11 207 kg. **6.** 0·00089; 1·502 $\pm$ 0·0017; significant $\therefore$ not consistent. **7.** 33. **8.** $\hat{\mu} = 20\cdot4$, $\hat{\sigma} = 0\cdot69$; $z = 1\cdot83$, not significant. **9.** $\bar{x} = 26\cdot1$, $S = 7\cdot67$; $26\cdot1\pm1\cdot14$; accept $\mu = 27$. **10.** $z = -1\cdot78$, significant; 111·8, 126·4; 1.

Exercise 13A, p. 187.

1. $t_9 = 1\cdot16$, not significant. **2.** $t_{13} = 2\cdot43$, significant. **3.** 2·17, 1·62; 2·17$\pm$0·60. **4.** $t_{14} = 1\cdot71$, not significant; 0·35$\pm$0·43. **5.** $\bar{x} = 3\cdot31$; $s = 0\cdot029$; 0·012; 3·31$\pm$0·03. **6.** 3·14$\pm$0·58; 3·15$\pm$0·31. **7.** $t_8 = 2\cdot40$, significant. **8.** $t_8 = -0\cdot26$, not significant. **9.** $z = 2\cdot1$, significant. **10.** 0·019; $t_{11} = -2\cdot70$, significant at 5%, not at 1%. **11.** $t_{98} = 2\cdot21$, significant at 5%, not at 1%. **12.** $t_{13} = 1\cdot87$, not significant. **13.** $z = 2\cdot71$, significant. **14.** $t_{12} = 0\cdot82$, not significant. **15.** (i) $\bar{x} = 32\cdot55$, $S = 4\cdot69$; (ii) $\bar{x} = 31\cdot67$, $S = 5\cdot10$, $z = 2\cdot05$, significant. **16.** $t_{11} = 2\cdot45$, significant; $t_{16} = 1\cdot15$, not significant. **18.** $z = 1\cdot48$, not significant. **19.** $z = 0\cdot99$, not significant. **20.** > 40.

Exercise 13B, p. 196.

1. $r = 2$, $P(r \leqslant 2) = \frac{29}{128}$, not significant. **2.** $r = 7$, $n = 8$, $P(r \geqslant 7) = \frac{9}{256}$, significant. **3.** $r = 5$, $n = 6$, $P(r \geqslant 5) = \frac{7}{64}$, not significant. **4.** $T = 6$, $P(T \leqslant 6) = \frac{13}{256}$, not significant. **5.** $z = -1\cdot27$, not significant. **6.** 8, $\frac{20}{3}$. **7.** $R_1 = 155$, $z = -1\cdot42$, not significant.

Exercise 14A, p. 204.

2. $\chi_9^2 = 17\cdot36$, significant. **4.** $p = \frac{1}{3}$; $\chi_4^2 = 4\cdot35$, not significant. **5.** $\bar{x} = S^2 = 3$; $\chi_3^2 = 2\cdot92$, not significant. **6.** Poisson good fit ($\chi_4^2 \simeq 2\cdot84$). **7.** $\bar{x} = 2\cdot38$; $\chi_5^2 = 7\cdot31$, not significant. **8.** $\chi_2^2 = 1\cdot19$, not significant. **9.** $\chi_4^2 = 1\cdot49$, not significant. **10.** $\chi_4^2 = 2\cdot71$, not significant.

Exercise 14B, p. 207.

1. $\chi_2^2 = 7\cdot11$, significant association. **3.** $\chi_1^2 = 9\cdot97$, significant association. **4.** $\chi_3^2 = 84\cdot75$, significant association. **5.** $\chi_1^2 = 3\cdot51$, not significant association; $\chi_1^2 = 2\cdot00$, not significant association; $\chi_1^2 = 5\cdot65$, significant association. **6.** $n > 40$. **7.** no; $\chi_1^2 = 3\cdot75$, $\chi_1^2 = 2\cdot68$, not significant association. **8.** $\chi_2^2 = 1\cdot54$, not significant association. **9.** $\binom{9}{2} \times \binom{9}{8}/\binom{16}{10} = \frac{27}{1144}$, $\binom{7}{1} \times \binom{9}{9}/\binom{16}{10} = \frac{1}{1144}$, significant association.

$$\frac{(a+b)!(c+d)!(a+c)!(b+d)!}{(a+b+c+d)!\, a!\, b!\, c!\, d!}.$$

Revision Examplos IV, p. 209.

1. Accept H_0 if $X \leqslant 2.9$; 0.4, accept H_0 if $X \leqslant 2\frac{2}{3}$.　**2.** (iii) $\frac{1}{40} = 2.5\%$.
3. $P = 0.0061$; (a) contradicted; (b) not contradicted.　　**4.** $P(r \leqslant 4) \simeq 0.029$, significant.　**5.** 7.9%; 8.5%.　　**6.** 11.　　**7.** $z = -1.25$, not significant; $z = -2.5$, significant; 250.　　**8.** $\bar{x} = 129.25$; $s = 13.8$; $s/\sqrt{n} = 0.87$; 129.25 ± 2.23.
9. ± 0.88 ohm.　　**10.** 1.44; 2.17.

11. $\mathrm{Var}(2\bar{x}) = \beta^2/3n$; $\mathrm{Var}\ \dfrac{n+1}{n}x_{(n)} = \dfrac{\beta^2}{n(n+2)}$

12. $\mu = 33\frac{1}{3}$, $\sigma = 4.71$; (a) 8 within 1 s.d.; (b) All within 2, supports hypothesis; $\bar{x} = 30.5$; $z \simeq 1.90$, accept randomization.　　**14.** $t_7 = 1.79$, not significant.
15. $t_9 = 2.66$, significant; $t_{16} = 0.99$, not significant.
16. $\bar{x} = 1.22$, $n = 64$; (i) not enough evidence; (ii) $z = 1.58$; evidence not significant.
17. $\chi_1^2 = 2.81$; not associated.　　**18.** $\chi_9^2 = 44.25$, significant.

Exercise 15A, p. 221.

1. -1.　**2.** 0.　**3.** 0.78.　**4.** 0.23.　**5.** -0.09.　**7.** points—goals against, -0.85; points—draws, -0.23.　**8.** 0.53.　**9.** 0.20.　**12.** 0.97.　**14.** $t_{12} = 2.0$; (i) reject; (ii) accept.　**15.** $t_6 = 0.59$; accept $\rho = 0$.　**16.** $t_{116} = 2.24$; reject $\rho = 0$.
17. 0.88; 0.63; 0.48; 0.42; 0.35; 0.25.

Exercise 15B, p. 225.

1. $r_s = 0.14$, $r_k = 0.07$.　**2.** $r_s = 0.38$, $r_k = 0.29$.　**4.** $r_s = 0.43$, $r_k = 0.20$; $r_s = 0.43$; $r_k = 0.20$.　**5.** $r_s = 0.91$, $r_k = 0.76$.　**6.** Points—goals for $r_s = 0.91$, $r_k = 0.76$; points—goals against $r_s = -0.86$, $r_k = -0.70$; points—draws $r_s = -0.19$, $r_k = -0.13$.　**7.** $r_s = -0.10$, $r_k = -0.083$.

Exercise 16, p. 235.

1. (i) $x = 28.4t + 176.8$; (ii) $y = 4.5t + 34.1$; x 375.6, 404.0; y 65.4, 69.9.　**2.** 203.4.
3. $863.4 + 54.5t$; 1299.　　**4.** 91, 83, 66.　　**5.** (a) 5.5, 11.92; 131.83, 48.14; (b) -0.25; (c) $x = 131.95 - 0.02t$; (d) -0.01.　　**7.** (i) $z = pbx + (q + pa)$; (ii) $y = a - ns/r + bw/r$.　**8.** 19.14, 15.03, $Y = 8.75 + 0.33X$.　　**9.** $\hat{m} = \Sigma xy / \Sigma x^2$.
10. $y = 0.98x^2 + 1.79$.　　**11.** $\hat{c} = 128.8$; $\hat{d} = 1.10$.　　**12.** $x = 0.99y + 1.48$; $x = -0.42z + 12.82$; $y = -0.32z + 10.82$; (a) $x = 10.69$, $y = 9.23$; (b) $x = 6.45$, $y = 6.05$; (c) $x = 10.66$; (d) $x = 7.50$.

Revision Examples V, p. 237.

2. 0.71; $t_{10} = 3.19$, significant.　**3.** $r_s = 0.77$; $r_k = 0.60$.　**4.** 103.0.　**5.** (a) 39.82; (b) $y = 51.70 + 0.40x$.　**6.** M 47, 19.39(S); P 58.75, 5.80(S): the mathematics mark; 31.1.　　**7.** $r = 0.23$; $r_s = 0.32$; $r_k = 0.20$.　　**8.** $x = 2$; $y = 0.51 + 0.50x$; 1.50; 0.0027.　　**9.** $y = \dfrac{211}{70} - \dfrac{39}{20}x + \dfrac{25}{28}x^2$.　　**10.** $\frac{1}{3}$.

Index

Index

Index